Collaborative Construction Information Management

Spon Research

Spon Research publishes a stream of advanced books for built environment researchers and professionals from one of the world's leading publishers.

Collaborative Construction Information Management

Edited by Geoffrey Qiping Shen, Peter Brandon and Andrew Baldwin

Spon Press
an imprint of Taylor & Francis

LONDON AND NEW YORK

First published 2009
by Spon Press

Published 2016 by Routledge
2 Park Square, Milton Park, Abingdon, Oxfordshire OX14 4RN
711 Third Avenue, New York, NY 10017

First issued in paperback 2016

Routledge is an imprint of the Taylor and Francis Group, an informa business

Typeset in Sabon by
Swales & Willis Ltd, Exeter, Devon

British Library Cataloguing in Publication Data
A catalogue record for this book is available from the British Library

Library of Congress Cataloging-in-Publication Data
Collaborative construction information management / edited by
Geoffrey Shen, Peter Brandon, and Andrew Baldwin.
        p. cm.
    Includes bibliographical references and index.
    1. Construction industry – Communication systems. 2. Information
resources management. 3. Expert systems (Computer science) 4.
Management information systems. I. Shen, Geoffrey. II. Brandon, P. S.
(Peter S.) III. Baldwin, A. N. (Andrew N.)
    TH215.C638 2009
    690.068 – dc22
    2008042797

ISBN 13: 978-1-138-99148-4 (pbk)
ISBN 13: 978-0-415-48422-0 (hbk)

Contents

Contributors

C. J. Anumba, Professor and Head, Department of Architectural Engineering, The Pennsylvania State University, USA.

Ghassan Aouad, Pro-Vice-Chancellor (Research), University of Salford, UK.

Godfried Augenbroe, College of Architecture, Georgia Institute of Technology, Atlanta, USA.

Simon Austin, Department of Civil and Building Engineering, Loughborough University, UK.

Matthew Bacon, FRSA, Chief Technology Officer, Integrated FM Limited, Warwick, UK.

Andrew Baldwin, Professor of Construction Management and Director of Innovative Manufacturing and Construction Research Centre (IMCRC), Loughborough University, UK.

Peter Brandon, Director of Salford University Think Lab and Director of Strategic Programmes, School of the Built Environment, University of Salford, UK.

Y. K. Chan, Construction Virtual Prototyping Laboratory, Department of Building and Real Estate, The Hong Kong Polytechnic University, Hong Kong.

Robin Drogemuller, Professor of Digital Design, Queensland University of Technology, Australia.

Shichao Fan, Department of Building and Real Estate, The Hong Kong Polytechnic University, Hong Kong.

Martin Fischer, Professor of Civil and Environmental Engineering, Director of the Center for Integrated Facility Engineering (CIFE), Stanford University, USA.

H. L. Guo, Construction Virtual Prototyping Laboratory, Department of Building and Real Estate, The Hong Kong Polytechnic University, Hong Kong.

T. Huang, Construction Virtual Prototyping Laboratory, Department of Building and Real Estate, The Hong Kong Polytechnic University, Hong Kong.

John Kelly, Director, Axoss Ltd, UK.

Arto Kiviniemi, Chair of the Salford Centre for Research and Innovation (SCRI) Steering Committee, University of Salford, UK, and a member of the 'Technology in Communities' Strategic Technology Steering Group, VTT, Finland.

Tuba Kocaturk, Programme Co-ordinator, Digital Architectural Design and Digital Design Initiatives, School of the Built Environment, University of Salford, UK.

Mohan Kumaraswamy, Executive Director, Centre for Infrastructure and Construction Industry Development, University of Hong Kong, Hong Kong.

Angela Lee, Programme Director, Architectural Design and Technology, School of the Built Environment, University of Salford, UK.

Heng Li, Chair Professor of Construction Informatics, Construction Virtual Prototyping Laboratory, Department of Building and Real Estate, The Hong Kong Polytechnic University, Hong Kong.

Steven Male, Professor of Construction Management, and Head of Construction Management Group, School of Civil Engineering, University of Leeds, UK.

E. A. Obonyo, Assistant Professor, M. E. Rinker, Sr. School of Building Construction, University of Florida, USA.

Rivka Oxman, School of the Built Environment, University of Salford, UK.

Martin Riese, Managing Director, Gehry Technologies Asia Ltd, Hong Kong.

Geoffrey Qiping Shen, Chair Professor of Construction Management, and Head of Department of Building and Real Estate, The Hong Kong Polytechnic University, Hong Kong.

M. Skitmore, School of Built Environment, Queensland University of Technology, Australia.

Souheil Soubra, Head of Division – Modelling and Enriched Virtual Environments, CSTB, France.

Kenny T. C. Tse, Department of Building and Real Estate, The Hong Kong Polytechnic University, Hong Kong.

Paul Waskett, ADePT Management Ltd, UK.

Jeffrey Wix, Consultant, AEC3 UK Ltd, UK.

Andy K. D. Wong, Associate Professor, Department of Building and Real Estate, The Hong Kong Polytechnic University, Hong Kong.

Francis K. W. Wong, Department of Building and Real Estate, The Hong Kong Polytechnic University, Hong Kong.

Song Wu, Lecturer, School of the Built Environment, University of Salford, UK.

Introduction

Collaborative construction information management – evolution and revolution

Andrew Baldwin, Geoffrey Qiping Shen and Peter Brandon

The aim of this book is to explore the way in which construction information management has the power to make the construction industry more effective and efficient. In particular it considers the growing interdependence between the participants in the process and their changing roles as they harness the power of the new technologies in new collaborative arrangements. It is not solely about technology, although this is the facilitator of change, but the manner in which behaviour will change as collaboration and communication are enhanced.

Introduction

The inefficiencies and low productivity of the construction industry are well documented. Despite the emergence and adoption of information and communication technologies to assist both design and construction the industry's productivity remains low. Compared to other industries, productivity in the US construction industry since 1964 has decreased (Li *et al.* 2008). Why? The construction industry is highly fragmented, it is dominated by small and medium-sized organisations, and in a typical construction project there are a large number of participant organisations. Communication and information exchange problems proliferate, and it is widely accepted that to overcome productivity problems construction organisations need to work collaboratively. How best can this be achieved?

In 1984, Irene Greif of MIT and Paul Cashman of Digital Equipment Cooperation organised a workshop attended by individuals interested in using technology to support people in their work. The workshop covered different industries. During this workshop, the term 'computer-supported cooperative work' (CSCW) was first coined to describe work supported by computer technology (Grudin 1994). Since then the word 'collaborative' has normally replaced 'cooperative', and 'computer-supported collaborative work' is the generally accepted term. This term recognises the importance of the computing, information and communication technologies to facilitate collaboration in construction.

Since 1984 there has been considerable research into the evolution of CSCW. This has focused on both the development of the supporting technologies and the human and organisational aspects of their adoption. Terms such as 'concurrent engineering' and 'groupware' have emerged together with others to describe collaborative working in a computer-supported working environment. This therefore is the topic of our book.

Collaboration is a highly complex and challenging task, which can be defined as 'The agreement among stakeholders to share their abilities in a particular process, and to achieve the objectives of the project as a whole' (Kalay 1998).

Collaborative working in construction means joint efforts among project stakeholders to effectively and efficiently accomplish a construction project. Collaborative working covers a spectrum of ways that two or more organisations can work together. Options range from informal networks, alliances or partnering to full integration. It can last for a fixed length of time or can form a permanent arrangement.

Collaborative construction information management represents a research area which focuses on a collaborative, integrative and multidisciplinary team of stakeholders tackling complex multi-scale issues involved in creating viable solutions in the context of the built environment.

Recent reports highlight the importance of collaborative working both now and in the future. It is argued that in the constantly changing global economy

> the ability to communicate over time and space, within and between organisations or communities, is essential to achieve this flexibility by making the best use of the knowledge and competencies available. Furthermore, collaborative environments are necessary to increase the productivity as well as the creativity by enabling new forms of work in production and knowledge intensive businesses.
> (European Commission Information Society and Media 2006)

In *The Future Workspaces*, a book providing insights into the findings of a major EU research project which looked at perspectives of mobile and collaborative working (Scaffers *et al.* 2006), the authors consider our places of work and reflect that these are in a transition and that 'the way in which we are organising our work has changed considerably over the past 20 years and will continue to do so in the future'. They provide a detailed vision regarding characteristics of mobility and future ways of working. This is summarised in Table I.1. Importantly they stress that this 'workplace innovation' 'is not just a matter of technology alone. Among the key issues are: workplace organisation, regulations, cross-organisational cooperation, management and leadership, organisational structure, business models, and incentive schemes.' The Future Workplace Study covers a range of industry

Table I.1 Mobility: traditional view and MOSAIC vision

	Traditional view	*MOSAIC vision*
Mobility concept	Mobility of the individual via forms such as teleworking and working on the move.	Mobility of both individuals and collaborating teams, work and workspaces.
Thrust	Support of mobile working within existing process boundaries.	Undertake work at the place and time according to need.
Values	Individualism, organisational efficiency and benefits.	Work–life balance, sustainable development, societal benefits.
Work environment	Individual workplace supporting people on the move at different locations in carrying out their work.	Collaborative virtual workspaces adapting to context and enabling work to be carried out anytime, anywhere.
Work	Work mostly oriented to communication activities and access to information.	Work oriented to all primary and secondary aspects of work and to collaborative working.

Source: Adapted from Scaffers *et al.* (2006).

environments including engineering. We share these visions and perspectives for collaborative working in construction.

We believe that the emergence of building information modelling (BIM), together with related technologies (such as virtual prototyping), represents a new way of working that will become widely adopted throughout the construction industry over the next 15–20 years. These new technologies, first introduced and subsequently developed by leading architects such as Frank Gehry, are now being embraced by numerous clients, designers and construction organisations worldwide. The adoption of this way of working and the benefits that may be accrued when it is adopted offer a new platform for information management and a new basis for collaborative working. These technologies have the potential to revolutionise the construction industry and enhance collaborative working. To achieve this, collaborative construction information management will be crucial.

This chapter reviews collaborative working and related technologies. It identifies the key aspects of BIM and related technologies currently available to facilitate collaborative working and highlights the contributions of the leading researchers and industry members who have contributed to this book. These contributors see BIM and related collaborative technologies as a platform for the emergence of a new form of collaboration that will inevitably lead to changes in the traditional roles of the participants within construction projects. Issues relating to information management are high-lighted. As with the emergence of all new technologies their rate of diffusion and adoption throughout the industry cannot be assured. The contributors also discuss inhibitors to change and how these may be overcome.

Collaborative working

Collaborative working in the modern context may be considered to have commenced with the advent of computing technologies. One of the origin-ators of the term 'computer-supported cooperative working' (CSCW), Irene Greif, commented that they coined the phrase partly as a shorthand way of referring to a set of concerns about supporting multiple individuals working together with computer systems. 'The meanings of individual words in the term were not especially highlighted' (Greif 1988). Since then different researchers have proposed alternatives to the word 'cooperative'. Coopera-tive working, or coordinated working, or collaborative working? What is clear is that even from the outset the focus was on 'presenting and discussing research and development achievements concerning the use of computer technologies to support collaborative activities, as well as the impact of digital collaboration technologies on users, groups, organizations and society' (Greif 1988).

Collaboration is widely interpreted as working in unison. Collaboration is more than working in an informal relationship or on a compatible mission; it denotes a more durable and persuasive relationship. Collaboration includes 'user communication and user awareness of each other's actions' (Shelbourn 2005). It is 'two or more companies working jointly to: share common information; plan their work content based on that shared infor-mation; and execute with greater success than when acting independently' (Barthelmess 2003). Collaboration may be seen as 'a philosophy of inter-action and personal lifestyle and cooperation as a structure of interaction designed to facilitate accomplishment of an end product or goal through people working together in groups' (Attaran and Attaran 2002).

In his study of collaborative working Grudin (1994) commented that CSCW started as an effort by technologists to learn from economists, social psychologists, anthropologists, organisational theorists, educators and any-one else who could shed light on group activity and that CSCW should contain two main concerns: the technology used to support people's work and how people work in this technology-supported environment. The authors concur and use the word 'collaboration' in all considerations, as we consider collaboration to include all aspects of technology, organisation, process and human factors, and emphasise the need for a holistic perspective.

Modes of collaborative working

The advent of modern computing technologies has resulted in different modes of collaborative working. Understanding and solving the main issues presented by collaborative working depend on conceptualising how people work (Palmer and Fields 1994). Anumba *et al.* (2002) describe modes of collaboration based on a classification of space and time. These are shown in Figure I.1. Typical forms of collaboration in space and time are also described by Attaran and Attaran (2002) and Baecker *et al.* (1995). The types of communication shown in Figure I.1 are discussed by Ugwu *et al.* (2001) and summarised below.

Face-to-face collaboration (synchronous mode) normally involves meeting in a common venue such as a meeting room, and participants engaging in face-to-face discussions. An example of this type of collaboration is a meeting between an architect and the structural engineers to discuss the implications of changes to the layout of a building. Face-to-face collaboration may also involve participants working at the same time in the same room on activities ranging from group decision making to group authoring or running a CAD program (see Palmer and Fields 1994).

Asynchronous collaboration means that activities take place at different times but in the same location. This mode of communication can be conducted using media such as notice or bulletin boards within an organisation.

Synchronous distributed collaboration involves activities that take place at the same time but where the participants are located at different sites. This involves real-time communication using one or any of a range of the current

	Same time	Different time
Same place	Face-to-face collaboration	Asynchronous collaboration
Different place	Synchronous distributed collaboration	Asynchronous distributed collaboration

Figure I.1 Collaboration models.

Source: See Anumba *et al.* (2002).

technologies and techniques such as telephones, computer-mediated conferencing, video conferencing, electronic group discussion or editing facilities and so on.

Asynchronous distributed collaboration is where activities take place at different sites at different times. This mode of communication involves communication via the post, for example periodic letters or news bulletins, fax machines, teletex, voicemail, pagers, electronic mail transmissions and so on.

Whatever mode of communication is adopted between the parties within the project it is essential that all parties are provided with updated, accurate information on which to base their decisions. Since its inception collaboration has been facilitated by different technologies. These technologies have changed as new technologies have emerged to supplement and replace those that already exist. This evolution is now examined.

Evolution

The last 25 years have seen considerable growth in the power and applicability of computers. This has dramatically increased the usefulness of digital electronics in nearly every segment of the world economy. Moore's Law, the name given to the findings of Gordon Moore, who noted that the number of transistors that can be inexpensively placed on an integrated circuit increased exponentially, doubling approximately every two years, describes the historical trend for computing power to increase exponentially (Moore 1965; Intel Corporation 2005). Similar increases in performance have been achieved with respect to computing performance per unit cost; power consumption; hard disk storage cost per unit of information; and so on.

Alongside this background of advances in computing power there have been considerable advances in software technology. The types of user interfaces between computers and system users have now consolidated, and de facto standards, for example for web-based systems, have emerged. East *et al.* (2004) conducted an in-depth analysis of the impact of web collaboration and conclude it is a very effective medium for conducting design reviews and offers many benefits over traditional manual methods of comment collection and resolution. There are a number of software developments available to facilitate collaborative working. Of all these applications, groupware demands special consideration. 'Groupware' is the name given to application software developed to support the collaboration of several users (Dix *et al.* 1998). Some researchers treat 'groupware' as another name for CSCW. However, others argue that it is incorrect to use 'groupware' and 'CSCW' synonymously because of the technological focus relating to groupware and the range of social forms developed within the context of cooperative working.

These technological advances have resulted in new ways of working for all industries. Adopting these technologies, some industries have radically changed their ways of working. Organisations have developed new business models to exploit the opportunities offered by the technologies. The construction industry has not however 're-engineered' its business processes to the same extent as many industries.

The range of hardware and software available for use on a typical construction project has been summarised and mapped in the process protocol (Kaqioglou *et al.* 1998). This map of the construction project process represents a basis for evaluating and establishing the process and the roles of the participants.

Effective collaborative working in service-based operations needs to bring together the four key resources of people, process, technology and data (Chapter 13). Within this book we cover all these aspects. These four resources are reviewed from the perspectives of both collaborative design and collaborative construction in a new era facilitated by the emergence of BIM and related technologies.

Revolution

Design for the built environment is seen as one of the most multidisciplinary practices in all of the design professions, since many professions, including architects, civil engineers, building services engineers, quantity surveyors, construction managers and landscape architects, are required to work closely during the design phase. Collaboration among different participants in the design of a building involves both synchronous and asynchronous communication. The different participants require the ability to work on their part of the project using their own particular ways of working yet being able to communicate with the other participants to bring about a common objective, the design of the building.

Three major and interrelated approaches towards collaborative working may be identified: technology and methods, business environment, and human behaviour. In this book, we have provided a number of showcases from each of the three approaches to demonstrate their applicability and success in addressing the unique problems existing in the construction industry.

Computer-aided systems have been used for the creation of site drawing and layouts for a long time (Tavakoli and Klika 1991). With the increase in computer-aided design (CAD) usage, there has been an increase in the interest in collaboration using the electronic medium. However, the use of CAD systems was primarily restricted to communication of shape information in the 1990s, much of it in the form of conventional drawings created on wire-frame-modelling-based systems (Veeramani *et al.* 1998). Such models are unable to meet the requirements of collaboration. With the emergence of new technologies such as the Internet and networking, things began to change.

A major UK research project, '3D to nD Modelling', funded by the Engineering and Physical Science Research Council (EPSRC) developed multidimensional computer models to portray and visually project the entire design and construction process, enabling users to 'see' and simulate the whole life of the project (Lee *et al.* 2003). An nD model incorporates multiaspects of design information required at each stage of the lifecycle of a facility, and provides a powerful mechanism to visualise the design, construction and operation process, and to integrate many other aspects of the process (Brandon *et al.* 2005). Marshall-Ponting and Aouad (2005) conclude that nD modelling could provide great value as a communication tool for industry and education.

In the aeronautical and automobile industries the use of BIM and VP is now commonplace, and all aspects of a new product or product changes are modelled virtually to assess the new product design, production and performance. This has revolutionised all aspects of design and delivery. We are currently experiencing a similar revolution in the design and construction of buildings. The last decade has seen the emergence of BIM and related technologies to a point where they may now be considered as the recognised platform for the design and construction of many construction projects. Their adoption has reached the 'tipping point' whereby their use may be expected to grow significantly over the next decade.

Eastman *et al.* (2008) define BIM as 'a modelling technology and associated set of procedures to produce, communicate and analyse building models' and characterise building models as intelligent digital representations of building components, which include data that describe how these behave. These data are consistent and non-redundant and may be combined in such a way that the model may be represented in a coordinated way.

Brandon and Kocaturk (2008) explore how BIM and the related technologies will present a new virtual future for design, construction and procurement. These technologies may be considered as a revolution for collaborative working because they will change *how* participants collaborate, *when* the project participants collaborate and the *contractual basis* under which they participate. The opportunities offered business organisations by these new technologies will result in the need for all businesses to review and refocus on how they add value to design and construction within this new business environment. They will also change the nature and timing of how design proceeds and how and when client value is added within the design and construction process. BIM and the related technologies will fundamentally change the project value chain.

BIM and related technologies will inevitably improve productivity and reduce waste within the construction process, change the role of professionals within the process and when and how they contribute their knowledge and expertise, enable data on individual projects to be shared within other larger models of the built environment, re-engineer existing business processes, and require new types of software and new technologies. Their introduction will succeed, however, only if the 'soft' aspects of systems and their implementation into organisations are considered, understood and taken into account.

Within this book the contributors review all these aspects.

Martin Riese (Chapter 8), in his review of the use of BIM on Swire Properties Hong Kong One Island East Tower project, describes how the adoption of BIM and related technologies throughout the design and construction process has helped to achieve a saving of at least 10 per cent in the cost of construction. He reviews the implementation of various key aspects of the building lifecycle information management techniques and the working methods that delivered success on the project.

The adoption of BIM will change the way that architects work. Kiviniemi and Fischer (Chapter 2) highlight that AEC practices are facing radical change because of the emergence of BIM. This change is affecting the architectural profession faster than other disciplines. They identify different dimensions of this problem and provide a practical approach to overcoming problems that emerge. Tuba Kocaturk (Chapter 14) identifies recent changes in architectural design culture due to extensive use of digital technologies and computational design environments, describes current digital design practice, and observes the ways and extent to which designers incorporate new tools and technologies into their working processes. She notes that

> One of the most crucial characteristics of this new field of design knowledge is that it is constructed collaboratively by the various parties taking part in the design and implementation processes. The emerging interactions between the design and production processes become highly non-linear and dynamic, leading to the emergence of a new, cross-disciplinary and collective body of design knowledge.

The importance of acquiring and utilising this design knowledge is emphasised by Rivka Oxman (Chapter 6), who looks at supporting multi-disciplinary collaboration through the integration of knowledge and information. She argues that 'computational enhancement of human collaboration ... demands a shift from information-based technologies to knowledge-based technologies' and that it is necessary to replace 'the concept of the "building information model" (BIM) with the concept of the "building knowledge model" (BKM) [which] may support such a shift in supporting human collaboration in digital design'.

Kenny T. C. Tse, Andy K. D. Wong and Francis K. W. Wong (Chapter 12) also use the One Island East project as a case study to examine the role of BIM and its impact on the professions, confirming that, whilst the role of professionals in a BIM environment will change, the importance of their domain knowledge will not.

In Chapter 5, Souheil Soubra 'explores the possibilities of using geospatial information as input data to construct 3D models of the built environment.

The 3D models are then combined with simulations in order to address sustainable urban development issues within the planning process.' He also reviews the social, organisational and human issues that need to be considered when working in an interdisciplinary manner.

Godfried Augenbroe (Chapter 11) looks at applying process rigour to the use of BIM in building design teams and reviews three technologies. He reminds us that

> Building design requires an orchestrated team effort in which many actors, tasks and activities have to be coordinated. As different actors use different software tools, each specialist traditionally operates on an island of isolation until the time comes to match and patch with other members of the design team.

This requires the ability to execute and manage a wide range of software applications. This challenge of seamless data interchange is probably the major barrier to the widespread adoption of building information modelling. Interoperability, the ability to seamlessly transfer data between applications software, is essential for success.

Achieving interoperability has been the challenge for many researchers and industry experts for a considerable time. The IFC ('Industry Foundation Classes' or, more descriptively, 'Information for Construction') schema defines a standardised file format that can be used as a mechanism for sharing semantically rich building information between CAD systems and an ever-expanding range of design analysis tools (Plume and Mitchell 2007). Fisher and Kam (2002) used IFC technology to facilitate data exchange among the major design partners in the project. They concluded that, compared to a conventional approach, these relatively seamless data exchange and technology tools substantially expedited design and improved the quality of interdisciplinary collaboration. Since IFC files are textural files whose size can reach 100 megabytes, Renaud *et al.* (2003) introduce an approach that can automatically identify business objects in IFC files and simplify their visualisation and manipulation on the Internet.

Robin Drogemuller discusses issues relating to the sharing of information between different disciplines in collaborative projects. The focus of his chapter (Chapter 3) is the information dependencies and the representation of the information required for collaborative working and how these issues may be addressed.

Jeffrey Wix (Chapter 10) considers that, as BIM applications become used more extensively and as object-based information exchange occurs, the capabilities and limitations of current developments are becoming more evident and new techniques are emerging that offer significantly greater potential to change the ways in which the building construction industry works. He argues for an approach based on an 'information delivery manual' (IDM), and describes a manual that has emerged as a response to the needs of software users in applying and trusting the IFC model for information exchange and also provides support to software developers.

As BIM develops, the technology and software to drive and support the process will change. Peter Brandon (Chapter 1) examines the impact of new technologies: access grids, tele-immersion, collaborative virtual environments and immersive collaborative environments. Ghassan Aouad, Song Wu and Angela Lee (Chapter 17) explore the opportunities to apply gaming technology into the construction arena. The computer game industry has invested heavily in the development of sophisticated game characters and achieved impressive advances in both technology and its user base. Such an approach for construction may be expected to deliver more complex, psychologically valid simulation models. E. A. Obonyo and C. J. Anumba (Chapter 7) also focus on information exchange and interoperability. It is their contention that, although significant strides have been made in refining the capabilities of such applications, there is still no overall integration scheme for the sharing of information between the existing tools. They propose the use of an agent-enhanced knowledge framework within virtual construction applications not only to address information integration issues but also to make the modelling more intuitive and hence powerful.

BIM and virtual reality (VR) produce the ability to model and visualise both the design and the construction process, thereby facilitating collaboration between all parties within the design and construction process. The term 'VR' is similar to and sometimes used synonymously with 'visual simulation', 'digital mock-up', 'virtual prototyping', 'walk-/flythrough' and '4D CAD' (Whyte *et al.* 1999). There are many VR applications in architectural design (Campbell 2000; Kolarevic *et al.* 2000; Caneparo 2001) and construction planning (Retik and Shapira 1999; Waly and Thabet 2003).

Rosenman *et al.* (2007) present a virtual environment framework for multidisciplinary collaborative design based on a virtual world platform and a model for representing the same design from the perspectives of different disciplines. It is proposed that the views of the various disciplines are modelled in separate hierarchies and the relationships between the various models are specified. Collaboration takes place in a virtual world environment because the multi-user and immersive properties of such environments facilitate synchronous communication and simultaneous modification to the different discipline designs. One of the main advantages of a virtual world environment is that it allows users to be immersed in the environment, allowing for real-time walkthroughs and collaboration (Conti *et al.* 2003; Savioja *et al.* 2003).

Heng Li, H. L. Guo, T. Huang, Y. K. Chan and M. Skitmore (Chapter 9) emphasise the need for change and the adoption of VR. They remind us of the low levels of efficiency consistently achieved in the construction industry and argue that to achieve greater productivity we need an 'IKEA approach' to management – designers working with manufacturers to find smart ways of production. This demands 'design without errors and appropriate construction sequencing'. This, they argue, can be achieved through the use of virtual prototyping (VP) to integrate design and production.

The process changes required to maximise the benefits of BIM and collaborative working are considered by Matthew Bacon (Chapter 13), who stresses that for service-based operations (such as construction) effective collaborative working will not be achieved through people working slavishly in their own professional disciplines, which will only serve to impede the process, but through the adoption of new roles that recognise the needs of new working practices. 'When people work together systematically using integrated processes, sharing common data, seamlessly exchanged between heterogeneous systems, an efficient and effective service is likely to be the outcome.' He then describes how this may be achieved. Andrew Baldwin, Simon Austin and Paul Waskett (Chapter 4) emphasise the benefits that can be accrued from modelling and managing the information flow within the design process and its importance in effecting collaborative working.

Experience shows that if we do not pay sufficient attention to the 'soft' aspects of systems implementation then new systems invariably fail. Mohan Kumaraswamy (Chapter 16) argues that rapid developments in hard system collaboration tools have overtaken the current capacities of most construction organisations and personnel to effectively mobilise, let alone rapidly optimise, such multi-dimensional management systems. He highlights international and local initiatives towards redressing the present imbalance between hard and soft systems, and addressing the growing gap in their future development.

Peter Brandon reminds us that the focus of the industry in the last decade has been the processes of design and construction and a more efficient and effective design and construction procurement whereby the number of interfaces have been reduced by technology to reduce the enormous overhead of communication. He focuses on the role of management in collaboration and catalysts for change and concludes that we are already seeing clients demanding that their professional teams use 3D models and that these in turn are expanding to provide a total knowledge structure for knowledge development. Improved collaborative environments may be achieved by minimising the interfaces through automation and improved collaborative working.

The overriding factor that will ensure the successful adoption of BIM within design and construction is increased client value. Eastman *et al.* (2008) argue that 'cost estimation integrated with a BIM design tool allows designers to carry out value engineering while they are designing, considering alternatives as they design, that make the best use of the client's resources.' Incremental value engineering while the project is being developed allows practical assessment throughout the design. 'BIM may therefore be expected to revolutionise Value Management and the collaborative working of parties to ensure that the construction client achieves maximum value from the new building' (Eastman *et al.* 2008). The contributors therefore also focus on this aspect of collaborative working.

Geoffrey Qiping Shen, Shichao Fan and John Kelly (Chapter 18) focus on communication in decision making within the field of value management (VM), which is one of the most widely used tools to harness the creative powers of a group of people to achieve more than the sum total of each contribution. By definition, VM is a function-oriented, systematic team approach to provide value in a product, system or service in which the decision is made corporately through collaborative working (SAVE International 1998). The process uses structured, team-oriented exercises that make explicit and appraise existing or generated solutions to a problem, by reference to the value requirement of the client (Male *et al.* 1998). As a result of technological development, uncertain economic conditions, social pressures and fierce competition, construction industry clients are placing increasing demands upon the industry in terms of the project quality, costs of delivery, time from inception to occupation and, above all, value for money of projects. VM, as a useful tool that can help the industry meet these challenges, has been widely used in many developed countries for several decades. The implementation of VM in a construction project is normally in the form of one or more workshops, which are attended by the major stakeholders, facilitated by a value specialist, and follow a 'systematic job plan'.

In Chapter 18 Geoffrey Qiping Shen, Shichao Fan and John Kelly introduce 'a group support system for collaborative working in a value management workshop environment' to aid the decision-making process. A group support system (GSS) is an interactive computer-based information system which combines the capabilities of communication technologies, database technologies, computer technologies and decision technologies to support the identification, analysis, formulation, evaluation and solution of semi-structured or unstructured problems by a group in a user-friendly computing environment. As there is a strong demand for improvements to the practice of VM, research has been conducted to design a GSS prototype system, named Interactive Value Management System (IVMS), to explore its potential application in VM workshops and to investigate the effect of the application. Chapter 18 begins with an introduction to the problems of implementing VM in the Hong Kong construction industry. It then provides an illustration of the features of the proposed group support system, which has been developed in the research. Two validation studies designed to test the support of the proposed system are described, and the results are discussed. Findings from this research indicate that IVMS is supportive in overcoming the problems and difficulties in VM workshops.

Steven Male extends the discussion on the use of new technologies for value management in Chapter 15, where he reviews the use of 3D computer visualisation methods in value management briefing and design studies. He presents case study vignettes of how the requirement for such methods has arisen and possible solutions, recognising that the use of such solutions is likely to be the domain of large-volume procuring clients, the large contractors and consultants.

Summary

Building information management and virtual prototyping enable new ways of working. They harness the power of technology to aid communication and thereby encourage collaboration. The book explores how various technologies, methods and approaches provide the catalyst for change and begin to change the nature and form of the design and construction process. The new technologies will change the roles of the participants, and it will be some time before this can be assessed, as it is an evolving process. As this form of collaborative working develops, other considerations will need to be considered, including the 'democratisation' of design, changes in the power of the community, and collaborative working in its widest sense whereby all the stakeholders are actively engaged in the design.

Whilst the benefits of BIM and related technologies are increasingly apparent, so too are the barriers to its adoption. These challenges include, but are not limited to, challenges with collaboration and teaming, changes in practice and use of information, implementation issues, and legal changes to documentation ownership and production (Eastman *et al.* 2008).

Despite the recent development, the construction industry remains fragmented; this is further complicated by the applications of isolated technical solutions and the lack of interoperability of design tools. The drive for improved collaboration includes the effectiveness of organisational operation, the need for more efficient use of resources, and the desire to accomplish more than through reductionist approaches where islands of sub-optimisation are developed. The aim is to make the whole greater than the sum of the parts.

There are enormous challenges ahead which will require collaboration between the participants to the construction project coupled with an efficient, effective information exchange developed through technology and improved processes. This book represents the views both of researchers and of academics working in the field. It is early days, but these insights will provide an indication of the direction the industry should follow. It is important that the discussion continues if the construction industry is to emerge as an efficient and effective force for the development of human activity and accommodation.

References

Anumba, C. J., Ugwu, O. O., Newnham, L. and Thorpe, A. (2002). 'Collaborative Design of Structures Using Intelligent Agents', *Automation in Construction*, 11(1), pp. 89–103.

Attaran, M. and Attaran, S. (2002). 'Collaborative Computing Technology: The Hot New Managing Tool', *Team Performance Management*, 8(1/2), pp. 13–20.

Baecker, R. M., Grudin, J., Buxton, W. A. S. and Greenberg, S. (1995). *Readings in Human–Computer Interaction: Towards the Year 2000*, 2nd edn, p. 742. San Francisco: Morgan Kaufmann Publishers.

Barthelmess, P. (2003). 'Collaboration and Coordination in Process-Centered Software Development Environments: A Review of the Literature', *Information and Software Technology*, 45, pp. 911–928.

Brandon, P. and Kocaturk, T. (eds) (2008). *Virtual Futures for Design, Construction and Procurement*. Oxford: Blackwell.

Brandon, P., Li, H. and Shen, Q. P. (2005). 'Construction IT and the "Tipping Point" ', *Automation in Construction*, 14(3), pp. 281–286.

Campbell, D. A. (2000). 'Architectural Construction Documents on the Web: VRML as a Case Study', *Automation in Construction*, 9(1), pp. 129–138.

Caneparo, L. (2001). 'Shared Virtual Reality for Design and Management: The Porta Susa Project', *Automation in Construction*, 10(2), pp. 217–228.

Conti, G., Ucelli, G. and De Amicis, R. (2003). 'JCAD-VR: A Multi-User Virtual Reality Design System for Conceptual Design', in *TOPICS: Reports of the INIGraphicsNet*, 15, pp. 7–9.

Dix, A., Finlay, J., Abowd, G. and Beale, R. (1998). *Human–Computer Interaction*, 2nd edn, p. 463. Harlow: Prentice Hall.

East, E. W., Kirby, J. G. and Perez, G. (2004). 'Improved Design Review through Web Collaboration', *Journal of Management in Engineering*, 20(2), pp. 51–55.

Eastman, C., Teicholz, P., Sacks, R. and Liston, K. (2008). *BIM Handbook: A Guide to Building Information Modeling*. Hoboken, NJ: Wiley.

European Commission Information Society and Media (2006). *New Collaborative Working Environments 2020: A Report on Industry-Led FP7 Consultations and 3rd Report of the Experts Group on Collaboration@work*. Brussels: European Commission.

Fisher, M. and Kam, C. (2002). CIFE Technical Report Number 143: PM4D Final Report. CIFE, Stanford University.

Greif, I. (1988). 'Remarks in Panel Discussion on "CSCW: What Does It Mean?" ', in *Proceedings of the Conference on Computer-Supported Cooperative Work*, 16–28 September, Portland, Oregon, ACM, New York.

Grudin, J. (1994). 'Computer-Supported Cooperative Work: History and Focus', *IEEE Computer*, 27(5), pp. 19–26.

Intel Corporation (2005). Excerpts from A Conversation with Gordon Moore: Moore's Law, ftp://download.intel.com/museum/Moores_Law/Video-Transcripts/Excepts_A_Conversation_with_Gordon_Moore.pdf (accessed via Wikipedia, Moore's Law).

Kalay, Y. E. (1998). 'P3: Computational Environment to Support Design Collaboration', *Automation in Construction*, 8(1), pp. 37–48.

Kaqioglou, M., Cooper, R., Aouad, G., Hinks, J. and Sexton, M. (1998). *Generic*

Design and Construction Process Protocol: Final Report. Salford: University of Salford.

Kolarevic, B., Schmitt, G., Hirschberg, U., Kurmann, D. and Johnson, B. (2000). 'An Experiment in Design Collaboration', *Automation in Construction*, 9(1), pp. 73–81.

Lee, A., Marshall-Ponting, A. and Aouad, G. (2003). 'Developing a Version of nD-enabled Construction', Construction IT Report, Construction IT Centre, Salford University.

Li, H., Huang, T., Kong, C. W., Guo, H. L., Baldwin, A. N., Chan, N. and Wong, J. (2008). 'Integrating Design and Construction through Virtual Prototyping', *Automation in Construction*, 17(8), November, pp. 915–922.

Male, S. P., Kelly, J. R., Fernie, S., Gronqvist, M. and Bowles, G. (1998). *Value Management Benchmark: A Good Practice Framework for Clients and Practitioners.* London: Thomas Telford.

Marshall-Ponting, A. J. and Aouad, G. (2005). 'An nD Modeling Approach to Improve Communication Processes for Construction', *Automation in Construction*, 14(3), pp. 311–321.

Moore, G. E. (1965). *Electronics*, 38(8), 19 April.

Palmer, J. D. and Fields, N. A. (1994). 'Computer-Supported Cooperative Work', *IEEE Computer*, 27(5), pp. 15–17.

Plume, J. and Mitchell, J. (2007). 'Collaborative Design Using a Shared IFC Building Model: Learning from Experience', *Automation in Construction*, 16(1), pp. 28–36.

Renaud, V., Christophe, C. and Christophe, N. (2003). 'Managing IFC for Civil Engineering Projects', in *Proceedings of the Twelfth International Conference on Information and Knowledge Management*, pp. 179–181.

Retik, A. and Shapira, A. (1999). 'VR-Based Planning of Construction Site Activities', *Automation in Construction*, 8(6), pp. 671–680.

Rosenman, M. A., Smith, G., Maher, M. L., Ding, L. and Marchant, D. (2007). 'Multidisciplinary collaborative design in virtual environments', *Automation in Construction*, 16(1), pp. 37–44.

SAVE International (1998). *Value Methodology Standard*, 2nd rev. edn. Northbrook, IL: SAVE International.

Savioja, L., Mantere, M., Olli, I., Ayravainen, S., Grohn, M. and Iso-Aho, J. (2003). 'Utilizing Virtual Environments in Construction Projects', *ITCon*, 8, pp. 85–99, http://www.itcon.org/cgi-bin/papers/Show?2003_7.

Scaffers, H., Brodt, T., Pallot, M. and Wolfgang, P. (2006). 'The Future Workspace – Perspectives on Mobile and Collaborative Working', a report for the MOSAIC Consortium Project funded by the European Commission as part of the Information Society Theme of the 7th Framework Research Fund.

Shelbourn, M., Bouchlaghem, N. M., Koseoglu, O. O. and Erdogan, B. (2005). 'Collaborative Working and Its Effect on the AEC Organisation', in *Proceedings of the 2005 ASCE International Conference on Computing in Civil Engineering*, Cancun, Mexico, 12–15 July.

Tavakoli, A. and Klika, K. L. (1991). 'Construction Management with AUTOCAD', *Journal of Management in Engineering*, 7(3), pp. 267–279.

Ugwu, O. O., Anumba, C. J. and Thorpe, A. (2001). 'Ontology Development for Agent-Based Collaborative Design', *Engineering Construction and Architectural Management*, 8(3), pp. 211–224.

Veeramani, D., Tserng, H. P. and Russell, J. S. (1998). 'Computer-Integrated Collaborative Design and Operation in the Construction Industry', *Automation in Construction*, 7(6), pp. 485–492.

Waly, A. F. and Thabet, W. Y. (2003). 'A Virtual Construction Environment for Preconstruction Planning', *Automation in Construction*, 12(2), pp. 139–154.

Whyte, J., Bouchlaghem, D. and Thorpe, T. (1999). 'Visualisation and Information: A Building Design Perspective', in *Proceedings of IEEE International Conference on Information Visualization*, London, 14 July 1999, pp. 104–109.

1 Collaboration

A technology or human interface problem?

Peter Brandon

The research agenda for construction has been dominated in recent years by the perceived need for a change in the *processes* of design and construction. In the UK, the Latham and Egan reports (Latham 1994; Egan 1998) have emphasized various aspects of this to enable a more efficient and effective design and construction procurement. The assumption has been that the process of construction design and assembly is where the focus should be, as this is where maximum benefit has been achieved in other industries. Part of this process issue has been the management of how the various parties to the process collaborate. If that collaboration breaks down then inefficiencies and abortive work emerge.

However, one of the major problems has been that traditional collaboration is operated through conventional models of procurement with a skilled but largely computer-naive workforce who are either unable or unwilling to adopt the new technologies to gain major long-term advantage. It is argued that the cost of change is just too high for an industry that works on low cost margins and would require a major restructuring of its workforce and a re-education programme for many of its employees. There have been no widespread role models which clearly demonstrate that any other alternative is superior and that commercial advantage can be achieved.

There is, perhaps, one major exception to the above and that is in the adoption of so-called *free form* architectures, because these structures cannot be built without the aid of 3D or nD models supported by a significant database of knowledge. In addition, the manufacture and assembly processes require a high level of technical support and tight tolerances which demand new forms of manufacture and CAD/CAM arrangements for major aspects of the construction superstructure. This necessity to harness the power of computers and their associated tooling mechanisms for manufacture provides a proving ground for the changes that might be expected for the wider industry.

At the moment this technological change is perceived to be costly, and there is concern that there are insufficient skilled workers to be able to use the technology effectively. The design team of most major projects is still operating only at the periphery of the technological revolution which could change the construction industry across the board. There is a strong need to reconsider the whole of the procurement process and, in particular, the processes of construction to establish what changes would be required to provide the most effective approach to collaborative design and manufacture to harness the new technological infrastructure which is emerging.

The changes required include education, the structure of the industry, the need for collaborative working across geographical boundaries, the models which support the new processes, the methods of communication, and the limits to innovation for construction. Free form architecture is currently the test bed for these activities, because the motivation is there, with a group of designers and technologists who are at the leading edge of both their own industry and that of the adoption of information and communication technologies (ICTs). There is a strong need to benchmark their progress against conventional approaches to demonstrate overall benefit but also to determine where the priorities for the revolutionary approach should lie.

This is not a trivial issue, and the following discussion seeks to highlight some of the matters that need to be considered, particularly with regard to the management of design and construction.

The role of management in collaboration

Process by its very nature is bound to consider the management of that process and where responsibility for activity lies. Management is an intervention which attempts to try to improve the efficiency and effectiveness of the activities which have to be undertaken to achieve a given end. Inevitably, this must include the manner in which all the various participants collaborate to reach the desired objective. Dictionary definitions of management are quite broad but include:

- 'to direct or control the use of';
- 'to exert control of';
- 'to direct or administer the affairs of';
- 'to contrive or arrange, succeed or accomplish, especially with difficulty'.

Much of the perceived difficulty lies in the way in which people collaborate and, increasingly, the way in which people use and interface with the technology.

However, there is a paradox in that as management is introduced to cope with complexity very often the more we manage the more complex the process becomes! Perhaps one of our targets should be to minimise the amount of management required in any given project without jeopardising the ability to be efficient and effective. Management is an overhead, and one of our targets should be to reduce it.

Of course, it is not possible to do entirely without management in some form. All we can do is question where it resides, who does what and how much it can be automated. The latter question suggests that it may be possible to reduce the human input at least, and this may well result in further efficiency. This is a role the technology can play, and innovative methods are now being developed.

The position of project management is already under debate from those who have had to challenge the traditional processes in order to get their buildings built. Often these are the designers who are responsible for some of the most exciting designs which are seen in our major cities. One of the pioneers in automating the process is Jim Glymph, previously senior partner of Frank Gehry Associates. His experience with dealing with the management of large-scale complex projects has led him to believe that in some instances project managers can get in the way.

In commenting on the Gehry fish sculpture for the Barcelona Olympics (Figure 1.1) he said that his firm agreed to go ahead with the project, which had a particularly tight schedule, if project managers were left out and there was agreement with the city of Barcelona that they could operate outside some of the conventions required by the authorities. These conventions

Figure 1.1 Barcelona fish sculpture (Frank O. Gehry & Associates Inc. (Gehry Partners, LLP), 1992).

and regulations were introduced to improve matters, but then the technology overtook them (e.g. the requirement for 2D drawings for regulation purposes when 3D is essential to understand a free form building) and they became impediments.

As Jim Glymph states with reference to the fish sculpture:

> In construction, you know, there's been a tradition, built up about paper and process, an approval process that is very complicated. We didn't sacrifice any quality control procedures; we clearly did not sacrifice any management; we just eliminated management where it was not necessary, which was most places.
>
> The fish sculpture was a fairly easy, steel structure, metal skin; it's not like the other buildings we are doing now ... but the big road block is still management.

Now this is a fine statement to make, but it is not quite as simple as it sounds. Because there was an element of automation introduced into the process then management was reduced, but what he is really arguing is for the designers to undertake the management in-house without a third party being involved and that the collaboration required to undertake the project is kept largely within the control of one organisation, thus avoiding the problems of collaboration between firms. This is a shift back to the processes adopted more than two centuries ago. Then the designer/engineer was responsible for all activity and management. With the rise of the general contractor and increasing complexity of buildings, more specialisms were introduced, resulting in more interfaces and knowledge silos, engaging many more professional advisers all trying to protect their own interests. In the end the designers (or their clients) outsourced nearly everything except form, specification, mass and space articulation on major projects (see Figure 1.2).

In more recent years, and particularly the last decade, the fragmentation has been identified as a major problem, and the introduction of new technologies which aid collaborative working, such as the internet, has suggested that we could move back to a united design/construction team aided and supported by technology, but it will mean a different set of procedures and protocols.

Addressing the problem

At the heart of the management function is dealing with the interfaces between people, organisations, physical artefacts, supply chains, technology and whatever else involves two or more people or artefacts creating some level of interdependence.

Figure 1.3 shows, conceptually, such a possible interface. It might be between two people or between people and technology or between a person

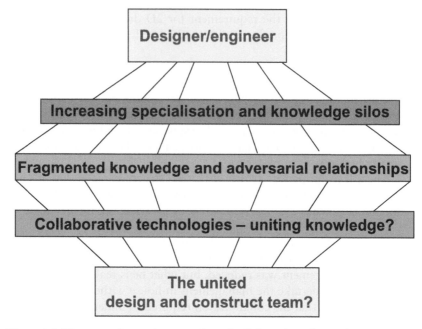

Figure 1.2 The expansion and contraction of collaborative relationships?

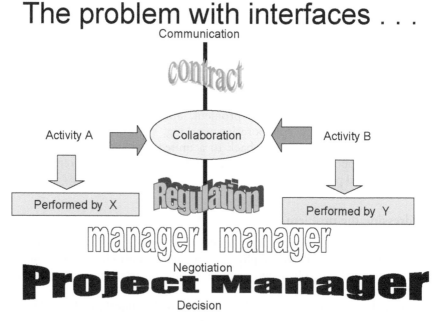

Figure 1.3 The issues surrounding collaborative interfaces.

and an organisation. It is worth noting the following possible build-up of the managerial function:

- Initially activity A needs to work with activity B.
- In order to do this both parties have to collaborate.
- The end result of the collaboration could involve a negotiation and/or a decision.
- The negotiation will involve some kind of communication, which is normally oral or written but could be electronic to a machine.
- At some point this process between the two may need to be clarified in a contract so that both parties understand the intentions of the other.
- The action and dependency might well be so significant that it is thought appropriate to develop regulatory powers to ensure that the public or the parties are protected in some way.
- The process is now becoming so complex that one party decides to appoint a specialist manager to handle the situation.
- The other party observes this and appoints a manager so that there is equal expertise available in the process.
- There comes a point when the managers can't agree and it needs someone to come in over the top and project-manage the managers! The project manager has arrived!

The above is, of course, a caricature of what happens, but nevertheless there is a ring of truth. In some circles this is called *creeping managerialism*, and it is something which pervades much of at least Western society, where the understanding of trust as an element of working practice has begun to break down.

The interface problem can be seen in a simplified cut-through of the procurement/estimating process. Figure 1.4 shows how information is transmitted through the system as the various parties attempt to collaborate.

In this case a set of professionals including the client are transmitting knowledge from one to the other to enable the building to be built. The process follows a familiar pattern:

- The client briefs the architect but can't articulate all his/her requirements – but does his/her best!
- The designer takes the brief and expands it from his/her own knowledge and experience and produces a design.
- The designer communicates the design through a model to the estimator, but the model is by definition a simplification of what he/she knows and is, therefore, not complete.
- The estimator interprets the model from his/her own experience and knowledge, expands some of the knowledge to suit the estimating process and prepares a bill of quantities.

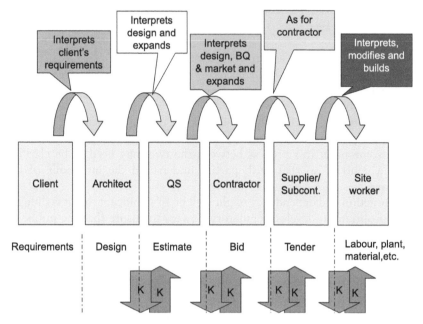

Figure 1.4 Professional interfaces and knowledge entropy.

- The bill of quantities is a simplified model of what is required to estimate cost, and therefore the bidders have to use their knowledge to expand the content and judge the cost of works.
- When they gain the job through competitive tender they then have to communicate to the site workers what has been assumed, and this again takes a different form and will be incomplete.
- The site worker takes the information and uses his/her expertise to actually make the information work in practice!

Throughout this process of collaboration, knowledge is being lost and gained. At each interface there may be a contract or some regulation that must be complied with and very often a negotiation/decision to be made based on incomplete information. The models used by the participants may not match each other. All these issues at the interface provide the potential for a breakdown, which can result in inefficiency and ineffectiveness. It is not clear that any amount of external management will solve the problem. At root is the interface itself, together with the models which each individual or organisation is using.

The result is:

- a failure in manual systems through potentially adversarial relationships;

- a failure of the IT systems used, as these tend to follow existing manual practice and support the interface position;
- a failure of innovation, because the systems are fossilized by the interface structure and the role of the professionals involved;
- creeping managerialism as more management is thrown at the problem.

Surely there is a better way!

Catalysts for change

Perhaps this is where the demands for new methods such as the structural components of free form architecture can play a part. These complex buildings have had to engage technology using three-dimensional and sometimes multi-dimensional models in order to understand, communicate, manipulate and manage processes. These are integrated models which have the power to revolutionise the industry. The technology is there and waiting to be used, and it is only the lack of conviction, knowledge and expertise in the rest of the industry which is holding back its implementation.

The introduction of 3D and building information management modelling systems has made some of the interfaces obsolete. If the whole of the design team are collaborating through a single model in which there is complete transparency and where the knowledge and intent of the collaborators are made explicit then the communication problems identified above can be largely overcome.

The DIVERCITY Project (DIVERCITY Handbook 2003) (Figure 1.5), developed by a group of leading research organisations between 2001 and 2003 under a European Framework programme, allows collaboration across international boundaries through a 3D model and information base. The result was a better understanding of each participant's role as well as better communication and interaction of those trying to collaborate.

There is still some way to go, but 3D modelling is now being demanded by some informed clients, backed up by the underlying knowledge base, in order to be able to use this data for the facilities management function and to avoid the costly delays in construction caused by incorrect communication. In particular, identification of clashes between components are thought to cost as much as 10 per cent of the project in delays and claims during construction.

Gehry Associates in the MIT Stata Center in Boston, MA (Figure 1.6) have taken the concept a stage further and have harnessed the power of CATIA (a software modelling system developed by Dassault in France) not only to model the building but also to link directly to the steel fabricator and suppliers of cladding panels so that these can be manufactured direct from the 3D model. Also related to the model is the ability to laser-scan physical models into the machine to create a virtual model so that the sculpturing of form can be done by the designer in the most appropriate way. The design

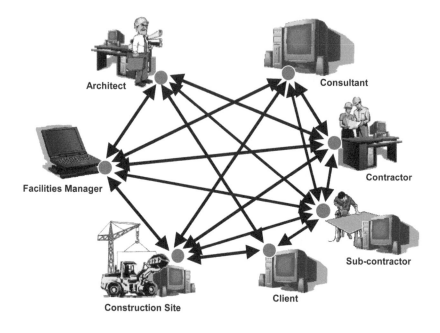

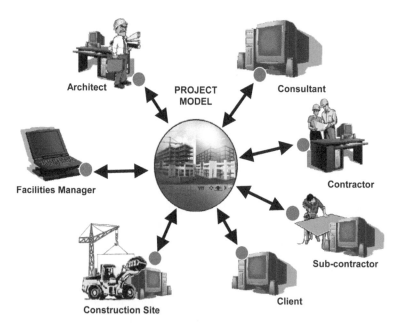

Figure 1.5 The Divercity Project: developing an integrated virtual prototyping tool.
Source: DIVERCITY Handbook (2003).

Figure 1.6 Rear elevation, Stata Center, MIT, Boston, MA (Frank O. Gehry &
Associates Inc. (Gehry Partners, LLP), 2004).

process then becomes a mixture of physical, mathematical, structural and
computer models each being used in a way which makes it most effective
and natural for the designer to use. This is then extended to the setting out of
the building on-site using lasers directed by the 3D computer model followed
by laser checks as the building progresses to ensure that it meets the tolerances
of the model. These checks identify where the actual build varies from the
computer model, and the manufacturing of the fabric is adjusted accordingly.
The variation is caused by the inability of manual labour to lay to the same
tolerances as the off-site manufacturer. Perhaps the next step is the robot
which can lay tiles and frame components to new levels of on-site tolerance.

The driving force for this innovation is the nature of many of Frank
Gehry's buildings. The rolling roofs and curved facades require a new
approach which can only be effectively implemented by the use of technology. Firms like Gehry are pioneers, and there are many problems to address,
as is the case with so many pioneering works in all fields. For example, the
project management workforce on the Stata Building at MIT, Boston, MA
was largely under 30 years old and had to be handpicked by Skanska to be
able to handle the technology. Employees above this age were thought to
carry too much baggage from the past. New forms of procurement had to
be developed which enabled the design/construction team to work in a

harmonious way, and the client was asked to take more of the risk than in more conventional projects. The number of suppliers/subcontractors able to work with this technology was limited, and this could create an uncompetitive market, and adversarial positions were minimised.

However, over time these potential hurdles will be overcome, and this approach will become the conventional approach to at least this type of building.

Collaboration between design and construction

At the present time this modelling concept is being driven still further with simulation of the whole construction process. It is in its early days, but Professor Heng Li at the Hong Kong Polytechnic University is using a 'bolt-on' package to CATIA called DELMIA, derived from the manufacturing industries, to simulate the construction process (see Figure 1.7). We are gradually moving to the point where we can build the building within the computer using various scenarios of resources and we can measure the difference in our methods before we even put a spade in the ground. Already, it is being used to examine the turn-round time for concrete frames/floors and the ability to use plant to maximum advantage. In one case it cut

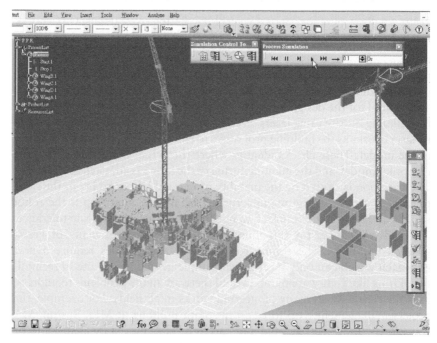

Figure 1.7 Simulation of construction process, Hong Kong.

Source: Heng Li, 2006 model: diagram generated from a public housing construction project in Hong Kong, Hong Kong Polytechnic University, Hong Kong.

the construct/pour cycle from six to four days per floor, and with high-rise blocks in Hong Kong this can be significant. At the moment, it is being sponsored by contractors who see immediate benefit and by clients who also feel that to some extent this benefit will be passed on to them. In this example the interfaces are transparent and can be dealt with at a time when the decision making is still fluid.

One further example of how free form helped to remove the interface is the simulation of the order of construction for the Chinese National Stadium in Beijing built for the 2008 Olympics (Figure 1.8).

Ove Arup, the consulting engineers, needed to think through the construction process as they designed and communicate this to the contractor. This was not an easy task to do, bearing in mind the complex 'bird's nest' nature of the design by Herzog & de Meuron. In this case they created a simulation model of several scenarios for a possible construction method in video form showing the order and method for the erection of the steel frame. Immediately it clarified the approach for a complex structure and made sure that all participants understood what was required. Again, it was the complexity of the form which pushed through the innovation. How else would the participants clearly understand?

The key question is whether these approaches can be translated to more conventional structures for equal benefit. There appears to be some

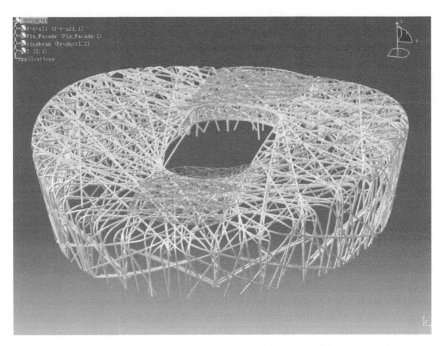

Figure 1.8 Chinese National Stadium, Beijing: early structural frame model.

Source: Herzog & de Meuron, Arup, and Chinese Architecture and Design Group.

evidence that this is, indeed, possible, and given time and some simplification of the technology, education of the workforce and willingness to adopt
then the construction industry can be transformed. For example, the work
done by Heng Li is being applied to fairly conventional multi-storey
buildings which would not, in any sense, be described as free form, yet
the benefits are clear to the clients concerned and in particular to Swire
Holdings, Hong Kong. They see real cash benefit in adopting the technologies and are now adopting them for all their major schemes. In addition,
the contractors who were required to use and develop the 3D model now
find it useful and are adopting the approach for other schemes.

Further underpinning work

By themselves, complex architectural forms do not introduce these changes,
nor does the technology answer all the issues involved in changing to a
more efficient and effective design/construct/operate environment for property and construction. It provides the *raison d'être* for adoption of the
technology but does not necessarily provide the integrated system which is
needed to challenge the status quo. A whole variety of tools is required
which sit within or on top of the 3D model to provide the maximum
beneficial effect.

One basic requirement is to understand the processes which go on during
design and construction and map them, partly to gain transparency but
partly because this mapping can provide a reference point for the project and
also identify where further work is required.

At Salford University, UK, a 'process protocol' (Cooper *et al.* 2005)
(Figure 1.9) has been developed in conjunction with industry which does just
this. Its origins are in the software industry, and it provides a list of events
without time scales which have to be undertaken to realise the project.

It consists of a matrix which has stages on the horizontal axis and activities on the vertical axis. At each stage it identifies what should be done at
various levels of detail. It is possible to drill down from each level to more
detail, and as you drill then the more bespoke the information becomes. At
the highest level the mapping is generic for most construction projects. Each
stage is concluded by a 'gate' which is either hard or soft. If it is hard then the
next stage cannot be implemented until the events have been concluded; if
it is soft then an assessment can be made but it is possible to progress.

If there is to be true collaboration then all parties must be aware of what
part they play in the design/construction/development process and where
their knowledge contributes to the overall objective. The process protocol
provides such a map and can be the point of reference for all concerned. It
makes explicit where the interfaces occur and which role player is involved.

Such a mapping would be useful in free form, and it would be interesting
to see whether the interfaces described within the protocol and its activities
are being reduced or eliminated. This could be a serious piece of research

Process protocol maps

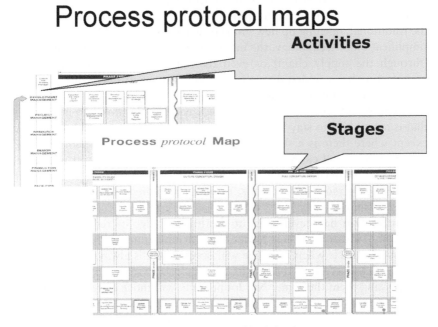

Figure 1.9 Salford University: process protocol high-level map.

which might open up still further the revolutionary approach to the industry behaviour.

Who is the motivator for change?

This question is one of the most difficult to answer within the traditional industry. However, in complex forms, at least up until now, it has largely been the client either directly or indirectly. Many clients select their architect on an ability to do something exciting, and this can often be free form when the building is expected to be iconic or at least a central feature of an organisation or community. Certainly, some architects will lean towards free form, and the client will know this.

By this selection the client is encouraging innovative design and by implication innovative practices which can involve innovative processes. Often the client takes the additional risk of veering away from the traditional processes.

The building and its processes thus become an experiment in new methods of design, procurement and construction. As time goes on and the team become more experienced then these innovative processes become the norm, and the envelope of what is possible is stretched still further.

The learning curve is not restricted just to the client or the architect, however. The people developing the software see new opportunities, and they begin to import them into their traditional packages. Knowledge begins to migrate in a similar fashion from all the participants from the special to the traditional, opening up new vistas of what is possible. This might be the simplification of technology, the integration of methods and understanding (through the supply chain) or even the nature of assembly, with more manufacture being undertaken off-site under more controlled conditions. In fact, the opportunities are enormous.

Of course, it may not be the client who generates the ability to innovate – that is most likely to stay with the professional consultants and contractors except where there is a strong in-house team of informed personnel. The driving force in all cases, however, is competitive advantage. It enables a designer to produce new ideas, a contractor to be more effective and an engineer to be more adventurous, resulting in more work and possibly a greater profit margin.

Construction has never been at the leading edge of research-sensitive industries! However, as these technologies demonstrate real business benefit then maybe this is when a change will occur. There is always the danger of viewing all the sectors of the construction industry as a homogeneous group and treating them the same. No doubt some will move faster than others. However, it is not easy to see what the long-term impediments are to a technological revolution in the industry, especially with the greater penetration of technology into all our lives. Grady Booch, the software guru and IBM chief scientist, in the annual Turing lecture in 2007 in Manchester, UK, suggested that the next decade would be about transparency in software development, the second would be dependence on the machine, and the third decade would see the rise of the machine. It would be interesting to plot what might happen to architecture and construction over such a time line.

The new technologies in tele-collaboration

The changes discussed above relate to current developments, but new technologies are emerging which provide new opportunities for collaboration to improve understanding and communication. The University of Salford, UK, 'Think Lab' has been designed to harness many of these new approaches, and they include:

- access grids where many participants can engage together through the technology and, allied to this, creative research to make the interface between people as natural as possible using eye contact and the ability to point at objects, and so on, to overcome some of the difficulties of videoconferencing;
- tracking technologies to allow movement to be captured so that this can be conveyed to an avatar or control device to ease communication or control;

- immersive devices whereby one or more people or objects can be immersed in the virtual environment and experience it in a similar way to real-life exposure. It is, therefore, possible to collaborate in cyberspace in a similar way to real life.

These technologies are still in their infancy and are improving and becoming more lifelike all the time. Figure 1.10 shows some of these technologies for collaboration:

- The access grid allows both parties to observe but not to share objects.
- Tele-immersion allows people to communicate as if they were at a desk and shared virtual objects.
- Collaborative virtual environments allow avatars to be controlled and interact with each other.
- Immersive collaborative environments allow individuals to 'meet' in a virtual environment and communicate.

Most research in this field is geared to making the interface more natural

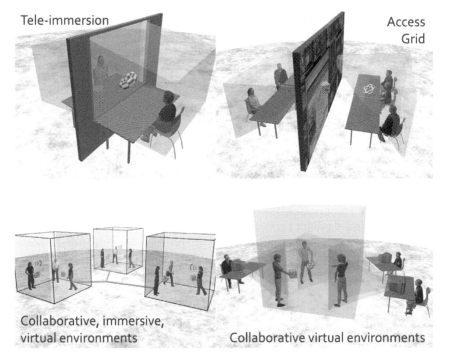

Figure 1.10 Different ways of sharing space in tele-collaboration.

Source: Professor David Roberts and the Centre for Virtual Environments, University of Salford.

so that collaboration is made more effective. In terms of construction the possibilities are enormous. For example, it may be possible to:

- have meetings in virtual space as naturally as if the people were geographically adjacent to each other;
- interact between people and the 3D models that have been created for design in real time but across geographical boundaries;
- control avatars which interact with the built environment to test safety issues in dangerous environments;
- examine and test in a safe environment, avoiding the potentially high cost of real-life testing;
- exchange objects across models to aid understanding.

In all these cases the collaboration is partly with the machine and partly with the human environment. However, as machines become more intelligent then it may be that the collaboration is between one machine and another. We are already seeing such collaboration between machines through knowledge grids where a number of machines work together calling on each other's knowledge to solve a problem. As the machine intelligence increases then the number of interfaces with human beings will fall and a less complex system will develop. However, the question remains as to how much we are prepared to, or indeed should, leave to the machine. It is important to realise that not everything we want to build in the machine can be built (as there exist pragmatic, theoretical and technical limits) and not everything we can build should be built (as there exist moral, economic, social and political limits).

Conclusion

The above discussion has focused on current trends and has suggested that free form architecture might well provide a role model for development for the whole of the construction industry.

Already we are seeing clients demanding that their professional teams use 3D models, and these, in turn, are expanding to provide a total knowledge structure for project development. Major clients such as the British Airports Authority are redesigning their processes to provide technological support and re-engineering of the supply chain to create a more efficient and effective process for procurement and construction.

Some contractors are learning, from working on large-scale contracts, new ways of using technology and are beginning to introduce the ideas into their traditional work. There is a domino effect as one innovation leads to another and the old practices begin to tumble down. Of course, we are not there yet, and many of the industry's practitioners remain sceptical as they gaze on the projects upon which they work which use some of these improved technologies. However, the pioneers are demonstrating and clearing the way, and the path is there for others to follow in due course. It is the second mouse that gets the cheese, and it may be easier for the second generation of design and construction personnel to enjoy the true benefits arising from this early work by the pioneers of digital architecture and manufacture.

To summarise, improved collaborative environments may be achieved by:

- minimising the interfaces through automation and improved collaborative working;
- maximising the technological adoption to aid design and manufacture as an integrated whole with fewer adversarial interfaces;
- monitoring and mapping more closely the processes adopted to examine and investigate where efficiencies can best be achieved by reducing interfaces through the engagement of technology to reduce or eliminate interfaces and improve understanding across all participants;
- motivating clients and each other to become the champions of the revolution for commercial advantage.

References

Booch, G. (2007). *The BCS/IET Manchester 2007 Annual Turing Lecture*, http:// intranet.cs.man.ac.uk/Events_subweb/special/turing07/ (accessed December 2007).

Cooper, R., Aouad, G., Lee, A., Wu, S., Kagioglou, M. and Fleming, A. (2005). *Process Management in Design and Construction*. Oxford: Blackwell Science.

DIVERCITY Handbook (2003). *The DIVERCITY Project: A Virtual Toolkit for Construction Briefing, Design and Management*. Manchester, UK: University of Salford.

Egan, J. (1998). *Rethinking Construction*, Report to Office of the Deputy Prime Minister, UK Government, http://www.asite.com/docs/rethinking_construction_ report.pdf (accessed November 2007).

Latham, M. (1994). *Constructing the Team: Report of the Government/Industry Review of Procurement and Contractual Arrangements in the UK Construction Industry*. London: Department of the Environment.

2 Potential obstacles to using BIM in architectural design

Arto Kiviniemi and Martin Fischer

Abstract

The architectural, engineering and construction (AEC) practices are facing a radical change because of the emerging use of building information models (BIM) instead of traditional CAD and 2D drafting. The change is starting to affect the work of architects faster than for other disciplines, mainly because architects must produce the basic model before the engineers and other construction professionals can efficiently use BIM. However, there are still also serious concerns, even scepticism, among architects about the usability of BIM as a creative design tool and about the effects the change may have in the professional practice and role of the architects in the new processes.

This chapter discusses briefly some dimensions of the problem; why is it difficult for many architects to move from drafting to 3D modelling although buildings are three-dimensional objects and spatial relations have paramount importance in architectural experience? Could it be that the background of architects' unwillingness to change their work process is what Heidegger calls 'readiness-to-hand' and what Merleau-Ponty calls 'intentional arc': our learned dispositions to respond to the situations instead of rule-based rational thinking?

The approach to the topic is mostly practical; the chapter discusses the issues from the viewpoint of building design using a mixture of literature review and phenomenological thinking, reflecting literature findings to the changes in the AEC industry, and also the first author's own professional background as an architect. The hypothesis is that architectural design process is not a rule-based rational system, but a complex mixture of intuition, skills, experience and personal and cultural values, and that this situation has significant impact in adopting new design tools and practices.

Introduction – brief history of design media

Our starting point is that the design medium influences strongly not only the visual expression but also the process and architecture (Hawkins 1997). The ancient practice of building design was based mostly on physical models and verbal instructions on-site, but for centuries drawings have been the main medium in architectural design. The development of formalized visual language improved communication among the participants, enabling more detailed documentation of the instructions as to what and how to build. This formalized visual language consists of several standardized 2D abstractions of a building, such as plans, sections, elevations, and isometric projections in different standardized scales, as well as perspective drawings.

For a long time the only copying method for drawings was manual tracing, which was naturally expensive. Thus the process was based mostly on unique drawings. However, this method was not limiting the representation, and the drawings were rather illustrative even for people who had no experience of how to interpret them (Figure 2.1).

The situation changed dramatically when the first copying machines were invented in the nineteenth century. The only possible representation for the early copying machines was black ink on transparent paper, and in the turn of the nineteenth and twentieth centuries the technology rapidly changed the visual language used in the building design.[1] The new representation was highly simplified abstraction compared to the earlier illustrative drawings, and, while the benefits of fast and cheap duplication were clear inside the process, the new language detached the process from the end-users, who had no education on how to interpret the more abstract drawings (Figure 2.2).

The technology 'privatized' the design language for professionals. One might even argue that it was also one of the bases for modernism, which

Figure 2.1 Old, illustrative elevation drawing: Merikasarmi, Carl Ludwig Engel, 1820.

Source: Museum of Finnish Architecture.

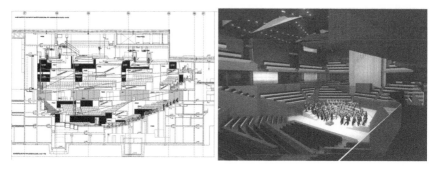

Figure 2.2 Typical modern drawing compared to the 3D visualization of the same space.

Source: Helsinki Music Centre, by courtesy of LPR Architects Ltd, 2008.

developed in 1920–1930, 'cleaning' the architecture from its traditional decorative elements and 'streamlining' the buildings. The phenomenon was of course affected by many parallel technical and social changes in society, and its roots were ideological (MIA 2008), but we can also assume that the new, more abstract representation most probably supported the change. Although this is just an assumption, the interaction between the media and representation is unavoidable, and the timely connection of these two phenomena supports our assumption.

The next revolution in design media happened in the early 1980s, when PCs and the first low-cost computer-aided design (CAD) software started to emerge in the AEC industry (Steele 2001). As usual in technical development, the new tools were imitating the old process, in this case automating the drafting process, instead of trying to take benefits from the new technological possibilities. However, some visionaries had already presented the idea of building product models earlier (Eastman 1975), and the first commercial low-cost 3D modelling tools for architectural design appeared on the market almost at the same time as the drafting software, but they gained only a limited market share, while 2D CAD became dominant.

When CAD started to emerge, several research efforts (van Leeuwen and van Zupthen 1994) were developing the ideas of 3D product models for the construction industry, and in the mid-1990s the idea of shared building models started to emerge among visionary practitioners, leading to the International Alliance for Interoperability (IAI) in 1996 (IAI 2008). IAI has been developing a product model standard called Industry Foundation Classes (IFC), which enables sharing of the BIM between different applications and direct utilization of the data for different purposes, for example energy simulations, quantity take-off, cost estimation, structural analysis and so on (Khalami 2004). The standard is today supported by all major software companies serving the AEC industry (IFC 2008), and several research projects and some pilot projects have demonstrated benefits of the shared model compared to the traditional drafting-based documentation (Froese 2002; Kam and Fischer 2002). However, the adoption of the new tools among practitioners has been slow. Though there are also several legal and contractual reasons for this, one reason is the resistance among professionals to adopting new design methods – even when there is evidence of the advantages.

What is architectural design?

Our view is that architectural design is a varying mixture of the three accounts presented by David Fallman (2003: 226–227): the conservative, the pragmatic and the romantic account. The degree of these different accounts in each case depends on the designer's personality, the nature of the building and the design phase/task. Some architects are more pragmatic, some are more artistic, some building projects have strong artistic expectations (like cultural monuments), some are highly practical (like simple warehouses or factories), and finally some design tasks are totally pragmatic (like door schedules) and some are mainly creative (like sketching). The whole building design process can hardly ever be a 'pure' example of any of these accounts, but it contains elements of all three; only the degree varies (Figure 2.3).

The building design process itself is based mostly on the conservative account; the process and its different stages as well as the roles and deliverables of the participants are well defined. Because of this the AEC companies are efficient in starting and managing new projects; everyone knows his/her role in the team immediately, and the effort to organize a new project is relatively small and easy; the problems are related mainly to the complex product, building, and the management of its information. This, combined with the degree of different accounts, determines which are the best tools and methods for each individual design task.

Acceptance of the new technology

The change from drafting on paper to the use of computers was technically a huge leap for architects. Before that, drafting equipment had been basically

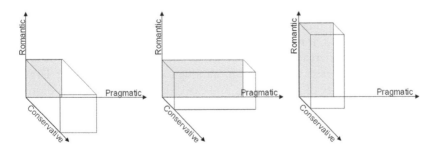

Figure 2.3 Different composites of three design accounts.

the same for about a hundred years. In addition, the pen is a perfect example of Heidegger's idea of readiness-to-hand. We have practised using the pen since we were children, and its use does not demand any conscious thinking. The user interface of a CAD program is complex, and achieving a sufficient skill level demands a lot of time and effort. Thus resistance in the beginning was very strong, and wide acceptance of the change took over a decade. However, the increasing productivity in the offices which adopted CAD early finally forced all architects to start using computers, and by the late 1990s 2D CAD had become a commodity in architectural and engineering offices. Today nobody questions the need for computers in design work.

However, the first step of CAD was mainly changing the drafting method, not the actual design process. With traditional CAD, architects still produce 2D drawings. Today digital models are used in many offices to some extent, but often just solid or surface models for visualization purposes, almost in the same way as perspective drawings were used in manual drafting. Only a few architects use BIM (AIA 2006) as the central source of information from which the necessary documents are generated, and even fewer exchange BIM with project partners.

The situation creates some obvious questions. For example: Is design using 3D modelling really different from that using 2D drafting and, if it is, how? If architects have already accepted the complexity of user interfaces in the CAD programs, why have they not accepted 3D modelling? Why have some architects accepted the change to 3D, but most have not? Is this change needed? Does it provide benefits, and to whom? What are the negative effects? To discuss these issues we divided the design process into two different phases, sketching and documentation. Their internal character is very different, and it affects both the process and the tools. A simplified description is that sketching is an internal thinking process and documentation is an external communication process.

The sketching process

Sketching is a crucial part of the creative process. As Black describes, 'Right from the earliest stages of tackling the problem, designers' thinking is mediated by the sketches or visible notes that they make to familiarize themselves with the material they are manipulating' (Black 1990). Sketching creates the basic ideas and, if we speak about architecture as an art form, the 'soul' of the building is created in the sketching phase. If this fails, the detailed design cannot correct the situation, although it naturally impacts largely on the final quality of the building and can either improve or even destroy the artistic ideas. In building design, the sketching process is also a deeply iterative process – a constant dialogue between ideas, analysis, synthesis and evaluation. 'It is indeed as much problem setting as problem solving' (Fallman 2003: 229).

In the sketching process the main purpose of the medium is to support the interaction between the designer's hand, eye and brain to develop the 'seed ideas' to something more concrete. It is a very delicate, even intimate, process. The 'language' is more or less personal, and usually one can, in the deep sense, understand early sketches made by another person only after a long collaboration, and even then the sketches are usually supported by verbal explanation of the ideas. Therefore sketches are often redrawn if they are intended for use outside of the core design team. In fact, some famous sketches in architectural books are drawn afterwards; they are not real working documents, but meant to support the image of the architect as an artist – although many architects may not be willing to admit this in public.

One additional important factor is that early sketches do not represent one 'classical' projection type; they are 'visual thinking', not 'formalized language' in the same sense as technical documents (Figure 2.4). Paper is still the only possible medium for most architects in this part of the process.

However, sketching on paper is inevitably limited to 2D projections; even perspective drawings are projections on a 2D surface. Therefore physical models have always been a central working medium for architects whose style is based on extremely complicated 3D geometry, like Frank O. Gehry (Figure 2.5). Gehry's innovation process is not based essentially on drawings, but on tens, in some cases hundreds, of physical working models, which serve as sketches which are digitized into computer models at some point of the process (Hajdri 2003). Gehry uses highly sophisticated computer technology in his work. However, the computer is not the main medium for his creative process, but an important tool to elaborate the ideas towards completion. The process is constant interaction between the physical and virtual models, and computers are an invaluable part of the creative process.[2]

Gehry's example is extreme in many ways; very few architects are able, and allowed, to design such buildings. Because of his success Gehry can also afford to use laborious methods like numerous physical models and expensive tools like CATIA (CATIA 2008), which would be financially

Figure 2.4 Architectural sketch.
Source: LPR Architects Ltd, 2008.

Figure 2.5 Walt Disney Concert Hall, Los Angeles (Frank O. Gehry, 2004).
Source: Arto Kiviniemi.

impossible for most architects. However, the geometry in most buildings is rather simple, and it is possible to sketch the forms using low-cost software in 3D, for example SketchUp®.[3] SketchUp® enables 3D sketching, and its user interface is very intuitive even with a mouse or pen tablets. When using an interactive pen screen the feel is very much like sketching on paper, only much more flexible because the designer can drag, push, pull and edit objects in the model. The model can also include hatching and real time shadows, and it can be exported to other software for the final design.

Whichever architectural style and design methods are used, to us it is clear that the creative process must be done using media which support the designer's immediate hand–eye–brain interaction best. Readiness-to-hand is an absolute demand; any need to think about the tool or its interface is a breakdown and disturbs concentration on the actual creation process. However, from our experience as teachers and the first author's experience as a designer, we know that the use of computer software can be learned on the level where it is an immediate extension of the designer's hand, and thus a useful medium also for sketching.

Nevertheless, it is necessary to ask if there really is a need to computerize sketching. Sketching is just a very small fraction of the architect's work measured by work hours,[4] and making it more efficient does not improve the whole process much even if we consider just the architect's own work. Considering the whole construction process, the efficiency of the sketching process means practically nothing. The quality of the design solution is much more important, and we do not believe that the use of a computer in sketching would make design solutions any better or worse. The quality comes deeper from the designer's skills, and the best tools are those supporting the designer's intent.

Design development and construction documentation

The architect's main efforts measured by work hours are design development and construction documentation. In addition, in those phases the need for communication increases dramatically compared to sketching because the number of project participants increases and the information includes more and more details, and both aspects create problems, which computer software can help to solve.

The representation in construction documents is highly regulated; the symbols for building elements, line weights, text sizes and other details are standardized for different drawing scales. Construction drawings are a formal language with their own syntax and semantics. Their main purpose is communication between the project shareholders. Drafting CAD applications were developed to produce these drawings, and they improved the productivity in design offices significantly compared to manual drafting; nobody can afford to work without CAD any more.

The logical question following from that is: Is it not enough? Why change to 3D? The answer becomes clear when one analyses the design and construction process in depth. Traditional documentation has one major problem, which cascades down to the construction process, causing even more problems downstream. The root of the problem lies in reducing a complex 3D object, the building, into multiple 2D representations, the drawings. This means that the same elements must be represented in several documents. As a simple example, the horizontal position of a window is presented in the plan, and its vertical position in at least one elevation and one section and often in several elevations and sections depending on the building geometry. In addition, there must be detailed drawings and verbal descriptions specifying a window's exact size, type, position in relation to the wall, fixing and caulking method, the type and colour of the finishes and glass, and so on. There are thousands of instances like that in one building, some 'owned'[5] by the architect, some by the mechanical, electrical, plumbing or structural engineer, and some by several participants, so that each 'owner' defines different properties of the element. For example, the colour of the ventilation and lighting fixtures is 'owned' by the architect, but the technical properties by the respective engineers. Of course all these details are also part of the design process, demanding decisions, not just recording existing data, but the process and problems are totally different compared to those of the sketching phase. The problem space in these details is more defined, and communication needs are much higher.

It is easy to understand that this complicated and fragmented documentation is a problem. Any change can affect tens of different documents, and there are hundreds or thousands of changes during a project. It is extremely difficult to remember or recognize all the documents which must be updated. We are willing to claim that it is in practice impossible; despite our long experience of the AEC processes, we have never heard of a project where the documentation was coherent. In the early 1990s a study in the UK (Latham 1994) concluded that information management is one of the main problems in the construction industry, and that it would be possible to save 30 per cent of the total construction costs if these problems could be solved. Information management is a real issue for the AEC industry.

Communication, information sharing and reuse

While the sketching process is 'internal communication' – a dialogue between the designer's brain, eyes and hand – the main issue in the design development and construction documentation is 'external communication' between project shareholders: the design team, owner, end-users of the building, cost estimators, contractors, suppliers and so on. The 'formal language' of construction documents has been developed to make this communication process efficient, and it is strongly based on drawings. Drawings have been the most efficient way to describe the necessary information; if we could 'write buildings' instead of drawing them, we certainly would have done so a long time ago. However, the medium sets its limitation to the message, and drawings are limited to the 2D space of paper.

Computer applications do not share this limitation, and as described earlier this gives new possibilities to manage the data through the BIM and solve the problem of fragmented information in multiple drawings and different representations. However, this does not necessarily mean that drawings would be obsolete: their role could change from the *storage of information* to a *view of the information*; the information is stored in the model and necessary drawings are generated from it.

At least in the current stage of technology this is a sensible approach. There are several reasons for this. The first one is the representation language itself; we do not have a tradition of how to read the models. In fact, BIMs contain information in such format and quantity that they are simply not human-readable; we must always look at them from some specific viewpoint related to the information we need in our task. Drawings are one of these viewpoints, and we are convinced that they will maintain their position among the views we will have of the BIMs in the future. There are some simple facts which support our claim. Drawings have been developed during a long period of time to minimize the transaction costs of the information exchange needed in the process. All AEC professionals understand the drawings immediately, and a simple drawing contains a huge amount of information which would be extremely difficult to present in any other format: locations, relationships between the building elements, measures, and so on.

Traditionally most drawings are produced to serve multiple purposes; the same plan is used in all the different tasks – design, cost estimation, form work, drywall installation, move management and so on – even though the necessary total information in each task is very different. This is an evident consequence from the paper-based process: the number of different documents must be minimized for two reasons: the direct work involved in their production and the amount of updates related to the changes described in the previous section. On the other hand, there are practical limitations on how much information one can add to one drawing and keep it readable. Together these aspects mean that the drawings must face trade-offs between their multi-usage and clarity. There are two practical consequences from this situation: the drawings include information which is not needed in some of the tasks they are used for, and some parts of the necessary information must be gathered from other drawings and documents; the information content of a drawing fits perfectly to any specific task only occasionally.

However, the new technology can improve even the drawings; if the drawing is only a temporary view of the information stored in the model, it can be generated to match with the task exactly, assuming that we know the information needs, and in any case significantly better than the traditional drawings. The 2D drafting CAD is already used in a 'primitive' way for this; it is fairly common that different layer compositions are plotted for a specific purpose on the construction site. A true BIM approach can improve this process significantly.

In addition, the possibility of producing different representations of the model goes beyond the traditional drawings, which are difficult to interpret for non-professionals and, in the case of complicated geometry, even for professionals. Different 3D representations can help people to understand the building on a totally different level. A good example of the difficulty in understanding the 2D drawings is that of one of the projects of the first author: Tapiola Cultural Centre in Finland (architect office Arto Sipinen, 1989) (Figure 2.6). The building has different floor heights in different parts of the building, and the only way to design the mechanical, electrical and piping (MEP) routes was to spend a whole week with the MEP and structural engineers in working sessions to define all the necessary details; in the usual buildings it is enough for the architect just to check their drawings.

An example of the use of a modern integrated model is the new Lucas Studio Project in San Francisco. There an external consultant built a detailed model based on the traditional construction drawings produced by the design team. Though the main reason for the model was its intended use for facility management purposes, it paid back more than the whole modelling effort by identifying several costly contradictions in the design documents early enough to correct them before construction work (Boryslawski 2004). In addition, according to the general contractor, the subcontractors often use the model to study different details of the building before the installation work, although their original attitude was that the model was just an 'expensive toy' (Davis 2004). The flexibility to choose the view in the model

Figure 2.6 Tapiola Cultural Centre (Arto Sipinen, 1989).
Source: Arto Kiviniemi.

is making a breakthrough among the professionals when they see the opportunities in practice.

The benefits of the model as the data source are not limited to the drawings; several tasks in the process are not based on the information related to the drawings, but on schedules and other written information, for example cost and quantity information. However, those are often strongly related to the building geometry; in fact, all quantity information is directly derived from it. Today in most cases the quantities are measured more or less manually from the drawings, and the process is both laborious and error prone. The possibility of linking cost, resource and time schedule information to the model is a significant process improvement from the cost and quality viewpoint (Kam and Fischer 2002; Agbas 2004; Fischer and Kunz 2004).

Another example of the achievable and even more significant benefits is the different building simulations. The most important of these is the energy simulation, because buildings use a significant proportion of the total energy consumption in all industrialized countries and thus the energy efficiency of buildings has major environmental and economic impact. For example, Lawrence Berkeley National Laboratory (LBNL) has over a period developed advanced simulation tools like DOE, DOE2 and EnergyPlus (Figure 2.7), but their usage in real construction projects has been very limited. The main reason for this has been the effort of creating a sufficient model of the building geometry. The possibility of importing the architect's BIM into the simulation software in the early project phases removes most of the modelling effort and improves the possibilities of studying different technical alternatives in the early design phases when it is still possible

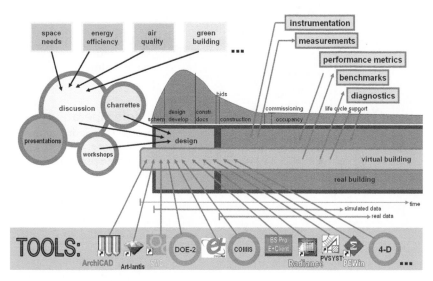

Figure 2.7 Some simulation tools.

Source: Vladimir Bazjanac, LBNL, 2003.

to try different solutions without significant additional costs (Figure 2.8) (Bazjanac 2003).

Likewise the benefits of the model are not limited to the design and construction phase, but the model also provides information for the facility use and maintenance databases, which are at present created manually after the completion of building, which is a similarly costly and error-prone process to, for example, cost estimation in the construction process.

As a conclusion, we argue that the integrated BIM information and its different representations compared to the fragmented information in the drawings are a significant improvement in all processes where project participants have to use and communicate complex, accurate and interrelated project information. The technology is available; the question is about learning how to use it in AEC practice.

Learning the visual language and working through it

Architectural education is based centrally on design projects. There are good reasons for this; one can learn to design only by designing. The process is very much as Dreyfus describes in his examples in *Intelligence without Representation*. It creates 'intentional arcs' through experience: 'skills which are "stored", not as representations in the mind, but as dispositions to respond to the solicitations of situations in the world' (Dreyfus 2002). In the design practice this means that an experienced designer is like a master-level chess player in Dreyfus's description, recognizing thousands of

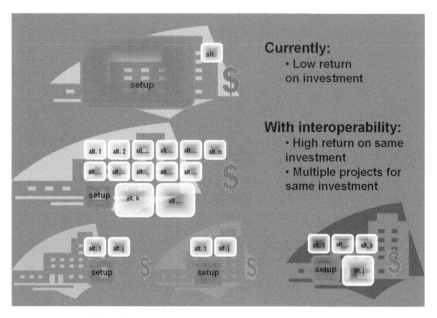

Figure 2.8 Modelling for the simulation versus use of the interoperable model.
Source: Vladimir Bazjanac, LBNL, 2003.

combinational situations instantly and identifying potentially usable solutions by intuition. The important role of intuition in the design process is emphasized not only by the famous architects but also in the formal definition of architectural education (UIA 1996).

However, the educational process in most universities is still strongly based on drawings – in some cases even manual drafting has a strong position in curricula. Even in the universities where modelling tools have an important role in education, they are in most cases used only for massing/ sculpturing and visualization purposes, as a 'modern version' of physical models and perspective drawings, not as a central source of information from which the production documents can be generated. To our knowledge, only a few faculties teach real interoperable BIM-based processes, simply because the paradigm is still new and most university teachers are not familiar with it.

Changes are slow because we transfer our culture to new generations. Designers do not just learn to *make drawings*; they learn to *think through drawings* (Fallman 2003: 229). We can relate the situation to Freeman's attractor theory describing the 'energy landscape' in our brains; it takes a lot of effort to change this 'energy landscape' after previous experiences have formed it. We believe that the 'intentional arc' and attractor theory can explain why it is so difficult for design experts to 'unlearn' previous practice. Learning any new design method and medium forces the expert back to a novice level in the skill acquisition phase, and it is understandable that most people do not want to take the step unless they are really forced to do it.

Development of design software

To us it seems that the development of traditional 2D CAD for architects was based on 'participatory design' (Winograd 1995: 118). To our knowledge, most CAD vendors had experienced architects in their product development teams, and they were mainly trying to make the drafting process more efficient, not considering if the medium had fundamentally different potential. Ehn poses two basic dilemmas (in Winograd 1995: 119): the 'gulf between the designer and user' and the tension between 'tradition and transcendence'. From this viewpoint the mainstream development was trying to narrow the gulf and maintain the tradition as much as possible.

However, Flores challenges that 'participatory design' would be the correct approach (Flores and Ludlow 1981), saying that 'design interventions create new ways of Being' (Winograd and Flores 1986). To us the difficulties 2D CAD had getting accepted in 1982 to 1995 seem to confirm this viewpoint; even a modest change in the medium was to some extent changing the process it was trying to improve. Another problem in 'participatory design' in our opinion is that it can provide only incremental improvements limited to the pre-understanding of the users. Using that method the full potential of the new medium is hard, if not impossible, to discover. New discoveries demand different thinking.

User interface

The user interface is a crucial part of the learning process; the less effort one needs to learn to understand the logic of new software, the easier it is to adopt. Standardization of the user interface in Windows has improved the situation compared to the early software running on MS-DOS; most software shares the same basic features – the idea copied from Apple. This helps the user in the learning process, and we do not think that the current user interfaces are the reason why modelling has not been adopted. The user interfaces in the best modelling software are not more complex than in the drafting CAD, but rather vice versa.

However, there is still a lot to improve in today's 3D CAD. Some totally new ideas, even tools, for modelling might come from the gaming industry, where immediate response and playability are mandatory for a product's survival. There are several features which modelling tools could adopt from the best games: great visual effects, speed, intuitive modelling tools for further development of the environment, and so on. Some research efforts to study how to use these possibilities for professional 3D CAD have been on-going for some time (Lehtinen 2002).

Shneiderman (1987) offers an analysis of the four stages of interaction:

1   *forming an intention:* internal mental characterization of a goal;
2   *selecting an action:* review possible actions and select most appropriate;
3   *executing the action:* carry out the action using the computer;
4   *evaluating the outcome:* check the results of executing the action.

Based on our own experience and also on Heidegger's and Dreyfus' ideas, we would challenge the validity of this model. The description fits to the novice level in the learning process as Dreyfus describes. However, for a master-level user, stages 1, 2 and 3 must become one on the conscious level; the user knows what he/she wants to do and does it. He/she does not select the actions by reviewing the possible alternatives and then execute the action. A good user interface must enable concentration on the content and forgetting the tool. The action must be like playing a musical instrument: never thinking about the commands, only the work one is doing. Shneiderman's stage 4, evaluation of the outcome, is a valid action, and it was already part of the process in drafting by hand: is the end-result satisfactory or do we want to change it?

Added value of the new technology – to whom?

In previous sections we have tried to show some evidence of the benefits of integrated BIM for the AEC industry and show that useful modelling software exists. However, integrated BIM is not widely used, and there must be reasons for that. If people can see sufficient benefits from the new technology, they will accept the change even if it is difficult. One reason why the benefits have not yet outweighed the difficulties is the financial incentive. The financial aspects are not a topic for this chapter, but they cannot be totally ignored. In a fragmented industry, such as AEC, the benefits of process improvements do not necessarily accrue to the company making the investments; architects' BIM gives major benefits to the other project participants, and many architects feel that they do not get sufficient compensation for the added value they would produce by sharing the models.

In addition, the medium can change the 'power structure' both within the AEC domain and inside an architect office. As described in the Introduction, the standardized 'language' for construction drawings is abstract and difficult to understand for non-professionals. Everyone who has worked as a designer knows that he/she has to explain the drawings to most clients. This creates an interesting situation, where the designer is interpreting his/her own work to the client. This gives a lot of factual decision power to the designer when there is a need to make choices between alternatives. It is easy to lead the client to the solution which the designer prefers. However, 3D visualizations change the situation; people are able to read the photorealistic images quite well. This means that the architect can lose the role of interpreter of his/her own work, and some, although hopefully not many, architects may see this as a disadvantage.

Another change in the 'power structure' can happen within the office when new tools are taken into use; several offices have experienced that the 'masters' of 2D CAD have more difficulties in learning the new 3D CAD compared to the people with less experience of 2D drafting CAD (Davis 2002). This supports the idea of 'unlearning' difficulties discussed earlier in the chapter, and in practice leads to a situation where the 'computer wizards' of the office may try to prevent the change. Skills and tools, like language, have politics too . . .

Conclusions

Software, like any new technology, must provide clear advantages to the users to be accepted. Whiteside writes: 'The purpose of contextual research methods is to uncover the user's experience of usability: that is, to identify dimensions of usability important to the user' (Whiteside *et al.* 1988). However, the benefits people experience are sometimes difficult to see in advance; there are examples of totally unexpected success in product development, like the Sony Walkman or SMS (Short Message Service); the latter was developed for GSM testing purposes, but it rapidly became the most profitable part of the mobile business in many countries. Value is a complex issue.

We do not believe that there are one or two simple reasons for the slow acceptance of 3D in architectural design. It is quite clear that most people are by nature cautious, even conservative. As a proverb says, 'better the devil we know'; new situations are for most people a threatening challenge, which they want to avoid as long as they can. This can also relate to the situational response; it is easier to act when the 'energy landscape' in our brain is familiar.

In addition, personal pride can be a big issue to some people; stepping down from a highly educated expert position to that of a novice learning new tools is not easy for most people. When we add these aspects to financial, contractual and legal issues and recognize that because of our mental structure the learning process will take a lot of energy from the productive design work, it is easier to understand why most people are not eager to change their professional practice.

The need to change the working method is a complex issue depending also on the environment, purpose and values. There is evidence of the benefits from modelling compared to drafting in the design development and documentation process, but, at least currently, not in the sketching process. To us the situation is somewhat similar to the difference between writing a dissertation and writing a tanka poem; a word processor is definitely a usable, even necessary, tool in creating a large document, but there are hardly real benefits with a short poem.

However, if we can understand the obstacles to the change, as well as the potential expectations of the users, it will help us to facilitate the change process and the design software environment which would be easier to accept and adopt. One of the conclusions is that the old design methods emphasized in many architectural schools might not just be futile but, by building the mental structures based on drawings, create real obstacles for architect students to learn new ways to think. Thus, the removal of technical drafting from curricula could be the right decision – though difficult to achieve in the near future because it may threaten the current power structure in the universities.

Notes

1 'In 1879 Silliman and Farnsworth of New York City were engaged as architects for the Chemical Laboratory Building of Vassar College. . . . Silliman brought from New York City three blueprints of the proposed laboratory building. . . . Over the next 20 years, a revolution seemed to sweep through American architectural and engineering offices. Tracing drawings to make copies was more or less eliminated' (Gustavson 2000).
2 'Much of what I have done in the last decade has been made feasible by our use of CATIA. I am excited by the prospect that we will be able to pass on our experience and know-how with computers to improve architecture overall. For me this would be a legacy more significant than my architectural designs' (Gehry 2003).
3 SketchUp® is commercial sketching software of Google. Starter edition is free, Pro version costs about $470 per license: http://www.sketchup.com/ (accessed 11 July 2008).
4 Conceptual design is typically 10–20 per cent of the architect's fee, and even in this phase sketching is only part of the work. In routine buildings, like standard dwellings or office buildings, the work sometimes starts based on some previous solution without any actual sketching. Thus, the sketching is normally 0–10 per cent of the total architectural design effort and, in most 'normal' buildings, less than 0.1 per cent of the construction cost.
5 In this context we mean by 'owning' the right to change the element in the design documentation.

References

Agbas, R. (2004). 'Geometry-Based Modeling and Simulation of Construction Processes', CIFE Technical Report 151, Stanford University, Stanford, CA.

American Institute of Architects (AIA) (2006). *The Business of Architecture*, pp. 43, 74, 75. Washington, DC: AIA.

Bazjanac, V. (2003). 'Virtual Building Environment: Connecting the Dots', American Institute of Architects Seminar, San Francisco.

Black, A. (1990). 'Visible Planning on Paper and on Screen', *Behaviour and Information Technology*, 9(4), pp. 283–296.

Boryslawski, M. (2004). Discussions with Mieczyslaw Boryslawski, View By View Inc.

CATIA (2008). 'CATIA is Dassault Systemes' Commercial PLM Solution for Digital Product Definition and Simulation', http://www.3ds.com/corporate/about-us/brands/catia/ (accessed 11 July 2008).

Davis, D. (2002). Discussions with Dianne Davis, AEC Infosystems, Inc.

Davis, G. (2004). Discussions with Gregg Davis, Webcor Inc.

Dreyfus, H. L. (2002). 'Intelligence without Representation: Merleau-Ponty's

Critique of Mental Representation', *Phenomenology and the Cognitive Sciences*, Special issue on 'Hubert Dreyfus and the Problem of Representation', ed. A. Jaap Jacobson, 1(4), Kluwer Academic Publishers.

Eastman, C. M. (1975). 'The Use of Computers instead of Drawings in Building Design', *Journal of the American Institute of Architects*, March, pp. 46–50.

Fallman, D. (2003). 'Design-Oriented Human–Computer Interaction', Proceedings of CHI2003, Conference on Human Factors in Computing Systems, Fort Lauderdale, FL, in *CHI Letters*, 5(1), pp. 226–227. New York: ACM Press.

Fischer, M. and Kunz, J. (2004). 'The Scope and Role of Information Technology in Construction', CIFE Technical Report TR156, Stanford University, Stanford, CA.

Flores, F.C. and Ludlow, J. (1981). 'Doing and Speaking in the Office', in G. Fick and R. Sprague (eds), *DSS: Issues and Challenges*. London: Pergamon Press.

Froese, T. (2002). 'Current Status and Future Trends of Model Based Interoperability', *eSM@rt 2002 Conference Proceedings Part A*, pp. 199–208. Salford: University of Salford.

Gehry, F. (2003). 'Gehry Technologies Extends Partnership with Dassault Systemes to Develop Solutions for Building Industry', http://findarticles.com/p/articles/mi_m0EIN/is_2003_Oct_23/ai_109155867 (accessed 11 July 2008).

Gustavson, D. M. (2000). 'Repro Roots', http://archiveink.com/HistoryofBlueprints.html (accessed 11 July 2008).

Hajdri, K. (2003). 'Bridging the Gap between Physical and Digital Models in Architectural Design Studios', *International Archives of the Photogrammetry, Remote Sensing and Spatial Information Sciences*, XXXIV-5/W10, p. 2.

Hawkins, D. C. (1997). 'Introduction Essay: Electronic Architecture', http://home.vicnet.net.au/~dchawk/invest1a/intro.htm (accessed 11 July 2008).

IAI (2008). International website, http://www.iai-international.org/ (accessed 11 July 2008).

IFC (2008). 'IFC Compliant Software', http://www.blis-project.org/BLIS_Product_Public.html and http://www.iai.hm.edu/ifc-compliant-software (accessed 11 July 2008).

Kam, C. and Fischer, M. (2002). 'PM4D Final Report', CIFE Technical Report 143, Stanford University, Stanford, CA.

Khalami, L. (2004). 'The IFC Building Model: A Look under the Hood', *AECbytes*, http://www.aecbytes.com/feature/IFCmodel.htm (accessed 11 July 2008).

Latham, M. (1994). *Constructing the Team: Report of the Government/Industry Review of Procurement and Contractual Arrangements in the UK Construction Industry*. London: Department of the Environment.

Lehtinen, S. (2002). 'Game Technologies: Previewing the Future of Modeling?', http://cic.vtt.fi/vera/Seminaarit/2002.04.24_IAI_Summit/Remedy.pdf (accessed 11 July 2008).

Minneapolis Institute of Arts (MIA) (2008). For example, De Stijl: http://www.artsmia.org/modernism/nintro.html, and Bauhaus: http://www.artsmia.org/modernism/iintro.html (accessed 11 July 2008).

Shneiderman, B. (1987). *Designing the User Interface: Strategies for Effective Human–Computer Interaction*. Reading, MA: Addison-Wesley.

Steele, J. (2001). 'Computers Have Revolutionized Architecture', quotation from K. Hajdri, 'Bridging the Gap between Physical and Digital Models in Architectural Design Studios', http://www.photogrammetry.ethz.ch/tarasp_workshop/papers/hadjri.pdf, p. 1 (accessed 11 July 2008).

UIA (1996). UIA/UNESCO Charter for Architectural Education, http://www.unesco.org/most/uiachart.htm (accessed 11 July 2008).

van Leeuwen, J. P. and van Zupthen, R. H. M. (1994). 'Architectural Product Modelling: A Case Study', Proceedings of the CIB W78 Workshop, Helsinki, http://www.ds.arch.tue.nl/Research/publications/jos/CIBW78_94/   (accessed 11 July 2008).

Whiteside, J., Bennett, J. and Holtzblatt, K. (1988). 'Usability Engineering: Our Experience and Evolution', in M. Helander (ed.), *Handbook of Human–Computer Interaction*, pp. 791–817. Amsterdam: North-Holland.

Winograd, T. (1995). 'Heidegger and the Design of Computer Systems', in Andrew Feenberg and Alastair Hannay (eds), *Technology and the Politics of Knowledge*. Bloomington: Indiana University Press.

Winograd, T. and Flores, F. C. (1986). *Understanding Computers and Cognition*, p. 163. Norwood, NJ: Ablex.

3 Collaboration using BIM

Results of Cooperative Research Centre for Construction Innovation projects

Robin Drogemuller

Abstract

This paper discusses the issues with sharing information between different disciplines in collaborative projects. The focus is on the information itself rather than the wider issues of collaboration. A range of projects carried out by the Cooperative Research Centre for Construction Innovation (CRC CI) in Australia is used to illustrate the issues.

Introduction

Within the CRC CI there has been a series of projects that used the Industry Foundation Classes (IFCs) as the basis for information exchange between various disciplines. While the projects were focused on supporting inter-operability between CAD software and the various software deliverables rather than purely on collaboration, a range of collaboration issues has been tackled within these projects. These issues have ranged from technical issues, such as supporting the appropriate flow(s) of information, to issues of responsibility, information quality and intellectual property.

There are many different aspects to collaboration and computer-supported collaborative work (van Leeuwen and van der Zee 2003). This chapter focuses on two of these – information dependencies and the representation of information. These two technical concepts underlie the level of confidence and trust that humans can have in collaboration activities.

While this is not leading-edge research within the context of collaboration, these results are important in identifying issues that need to be tackled in the short term by current software to support collaboration in the architecture/engineering/construction facilities management (AEC-FM) industry. The contribution of this chapter is to show how these issues can be addressed within software tools that are either being commercialised or were developed with commercialisation in mind.

CRC CI-sponsored IFC-based projects

The CRC CI has sponsored a series of projects over the last six years that cover most of the project lifecycle and a range of functions within the design/construction/facility management team. The project deliverables that will be considered in this chapter are set out in Table 3.1, showing their respective positions within the standard building project lifecycle.

A number of the deliverables support multiple stages within a project. This is achieved through consideration of the ways that information is gradually added during the building project. This highlights one of the most important issues when considering collaboration – the quantity, quality and level of detail of available information gradually increases throughout the project until the building is initially completed. If the building manager is sufficiently motivated and the information is available in an appropriate format then this information can be maintained throughout the operation of the building and can then be used when the building is being refurbished or demolished to provide accurate information to the project team.

IFC model

The IFC model was used as the data model for these projects, as it is supported by a range of commercial software, and the intention within the CRC CI was that deliverables from projects should have the potential for commercialisation if that was perceived to be appropriate.

Table 3.1 Position of CRC for construction innovation deliverables in the procurement process

Early design	Detailed design	Pre-construction	Construction	Facility management	Re-lifing
Automated estimator			*Automated scheduler*	*Integrated FM*	*Regenerating construction*
Parametrics for massing studies	LCA design				
Microclimates	Design spec				
Design check					
Area calculations for NS3940					
Design view					

The IFC model (IAI 2006) is defined in the Express language and bears a strong similarity to the STEP (ISO 10303) standards. Since it is defined using Express, the definitions are object-based, consisting of an entity (object) type, possibly a parent entity and a list of attributes, with some being inherited from the parent entity. The IFC model uses only single inheritance even though Express supports multiple inheritance. The aim of this decision was to simplify implementation.

The IFC model was not intended to cover the entire design and construction process right from the start, so a number of features were added to support extensibility. The scope of the formally defined part of the model has been growing on a regular basis since the first release.

The concept of 'proxy' objects was used to allow the dynamic extension of the range of object types. Hence new objects can be added to support a particular information exchange scenario even if the current IFC specification does not explicitly support the objects. The concept of property sets is also used within the IFC model to extend the number and characteristics of attributes within an entity or instance within an exchange scenario.

CRC CI deliverables and support for collaboration

The following sections cover software deliverables from the CRC CI where they highlight particular aspects of computer-supported collaboration.

A major issue that has occurred across all projects is the treatment of 'proxy' objects. Where possible, proxy objects are mapped to the appropriate IFC entity if it exists. This was very common in the earlier implementations of the IFC where library objects were automatically treated as proxy objects even if there was a straight mapping to an existing IFC entity. Currently, this is mainly an issue when users have not filled the relevant fields properly.

Another generic issue is where a necessary building component is not defined within the IFC model. For example, in the versions prior to IFC 2×2 there was no concept of a (foundation) pier. Rules were defined to allow mappings between building components that were supported to those that were not so that then current CAD software could be used. One rule stated that if a concrete column existed that was below the lowest slab and its base was not connected to anything then it was to be converted to a pier.

Design view

Design View is the central component within the ICT implementation strategy within the CRC CI (Figure 3.1). It is built on top of the Eclipse software development platform (Eclipse 2007a, 2007b; Eclipse.org 2008). It provides several functions that are shared across the various software deliverables. The major function is visualisation, with a 3D viewer, a tree view of the

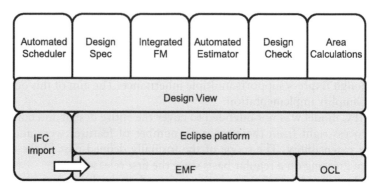

Figure 3.1 Plug-ins to Design View.

object hierarchy within the data model and a property view of the currently selected object.

Underlying the views is a shared data model that consists of a minimal union of the information needs of the various plug-ins that have been written. This data model is implemented using the Eclipse Modeling Framework (EMF) (Eclipse Modeling Framework Project 2008), an object-oriented representation that is integrated with UML. EMF also supports the W3C Object Constraint Language (OCL) (OMG 2006), which provides rule checking and mapping facilities.

A facility that is inherited from the Eclipse platform is a 'Tasks' list and a 'Problems' list (Figure 3.2). These provide rudimentary support for data quality. The 'Tasks' list allows someone to record the issues that they are aware of that still need to be addressed. For example, this allows an architect to indicate 'I know that the finishes are not yet added to the model. For estimation purposes assume X.' The 'Problems' list allows automated checking of an incoming model to ensure that all of the expected data is there. Currently the checking is limited to checking that all of the required attributes have an appropriate values set. If information quality is to be more fully supported there will obviously need to be integration between the 'Tasks' and 'Problem' lists as well as formal agreement between the sending and receiving parties as to the required information at particular types in the project lifecycle.

One of the first collaboration issues that arose in the CRC CI projects was the need to provide visualisation that was integrated with the various analysis tools. This provided the following functions:

• means of browsing the results and associating data with location and spatial relationships;
• methods of visually partitioning the model to examine particular types of objects or systems;

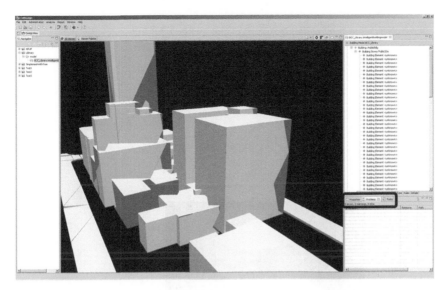

Figure 3.2 Design View user interface.

- checking of the model and an indication of objects that were not 'understood';
- the ability to examine and edit some properties of objects.

These all played a critical role in allowing the users to trust both the software and the data they were receiving. The integration of analysis and visualisation was considered so useful that a second version of the IFC viewer was planned. This, with some additional features, became Design View.

Automated estimator

The Automated Estimator reads the BIM model and automatically takes off the quantities that are defined for each trade. The quantities may be simple linear, area or volume calculations on a single component or they may aggregate a number of components under the one item, such as a concrete slab with attached beams and thickenings (Figure 3.3).

In some instances, components defined in an appropriate manner for one discipline are aggregated for another discipline. For example, in architectural CAD a column, beams and a slab may be defined as distinct components. However, for the concretor and the form worker they will be considered as only two components (Figure 3.4). Care is obviously needed to avoid double counting when components of an assembly and the assembly itself are stored in the same database. This is currently handled by explicitly tagging components for each trade package within the bill of quantities (BoQ).

Each item in the BoQ consists of a number of components. The textual description is stored independently of the database query that extracts and

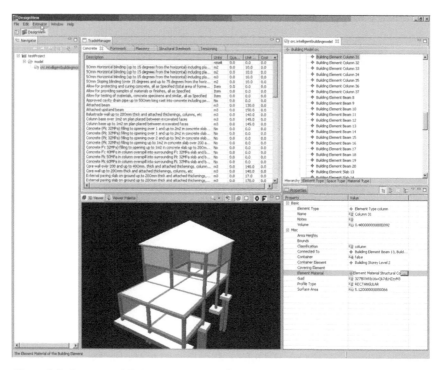

Figure 3.3 Automated Estimator user interface.

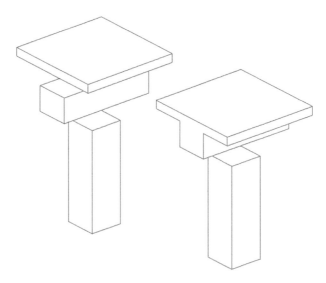

Figure 3.4 Architectural (left) and quantity take-off (right) views (exploded).

collates the information. This allows a level of natural language independence but is not as sophisticated as in the 'Area Calculations' work described later.

Design check

Design Check supports the automated checking of BIM models against codes and standards. It has the important differentiation of allowing the encoding of an entire code and the results for each clause through a single interface, whether the BIM model is used to verify the clause or not.

The range of possible results from a checking run is shown in Figure 3.5. The green ticks indicate that enough information was available within the BIM model. The other symbols mean that there is user input required before an issue can be resolved.

Figure 3.5 Design Check clause selection user interface.

Figure 3.6 Design Check results view.

The type of detailed output is shown in Figure 3.6. As the tabs indicate, the user can add explanatory information to the model.

One issue that did not arise in development and testing but soon became obvious in use on real projects was the need to structure the output so that people were not overwhelmed with data. The first time that the software was used on a live project (Figure 3.7) it generated over 500 pages of issues. Most of these issues were the same problem applied to many objects, such as minor problems with door widths, or a multitude of problems with one object, such as a particular room.

Most of the identified issues would have been expected in a project at that stage of development, so the reports did not add too much to collaboration between the designer and code checker. The main issue here is that results should be presented as concisely as possible and at an appropriate level of detail for the stage of the project.

Parametrics for massing studies

The most ambitious project undertaken within the CRC CI from a collaboration perspective was the 'Parametric Building Development during Early Design' project. This project examined the information dependencies across the architectural, cost and engineering (structural, hydraulics, mechanical

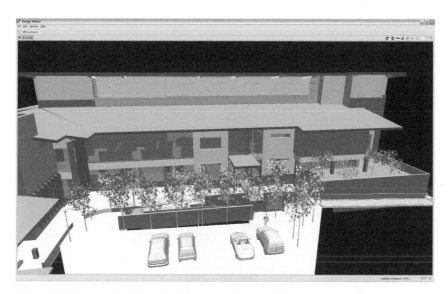

Figure 3.7 Queensland State Archives building model.

and electrical) disciplines during the early stages of design. The algorithms used were based on Parlour (1997).

The aim was to demonstrate that complex information interdependencies could be handled at the early stages of design. The software deliverable was a proof of concept.

The input to the system was a simple massing model (Figure 3.8) of the proposed multi-storey building with one occupancy type per floor. In Figure 3.8 the shading represents the occupancies – lightest is residential, medium is commercial and dark is below-ground car parking. The services core was near the centre of the model.

The key decision points within each architectural and engineering system were then defined (Figure 3.9). The interrelationships between each system were then identified and the dependencies established. These were then placed in a graph diagram (Figure 3.10). Each node represents a decision, with the input and output data for each decision represented by the database icon. The network forms a monotonic reasoning chain using a method called Perspectors, defined by Haymaker (Haymaker *et al.* 2004). This means that if a decision on the left-hand side of the diagram is modified every decision connected to it to the right has to be checked for validity.

The following chain is an example of the flow of decisions in the network. Increase the floor-to-floor height from 3.6 metres to 4.0 metres. This increases the external area of the building subject to wind loads. The depth of the beams in the floor needs to increase by 0.1 metre to increase the stiffness. Thus the floor-to-floor height becomes 4.1 metres. Since this is still structurally acceptable this height can stay. Now check the total height of

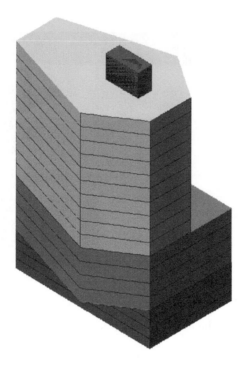

Figure 3.8 Massing model.

the lift (elevator) shafts. Is the speed of the lift acceptable for the building type and standard? If not, either increase the capacity of the lift cars or add another lift shaft. If another lift shaft is added then the total floor area is increased, and so on.

The design parameters for each of the architectural/engineering systems is specified in a diagram similar to Figure 3.10. Any available input variables are editable by the user. Dependent variables are automatically updated for the system under evaluation and other dependent systems.

Area calculations for NS3940

This project involved developing a plug-in for Design View and calculated four different areas as specified in the Norwegian standard NS3940 (Figure 3.11).

An obvious issue with this project was the use of natural language to exchange information within the IFC model. For example, 'betong', the Norwegian term for concrete, is used in the material description in the IFC file.

A major issue with the use of terms is that, within a single language, the same word may be used in several different ways. For example, a 'door' may

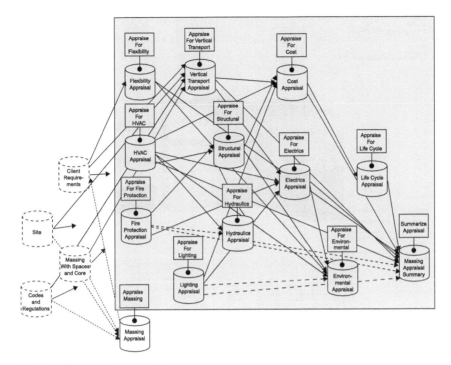

Figure 3.9 Decision network.

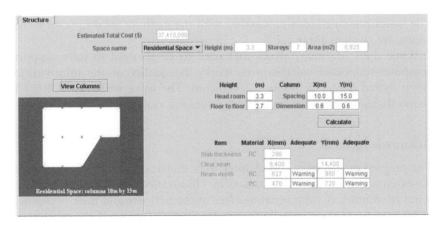

Figure 3.10 System parameters dialogue.

Figure 3.11   Area Calculations user interface.

refer to the 'door leaf', 'door leaf and door frame' or 'door, frame and hardware'. This issue is being addressed through the Lexicon (STABU) and IFD (http://www.ifd-library.org/) projects.

Support for collaboration

The issues presented above, together with the methods used to resolve the issues, are given in Table 3.2.

While there are many issues that need to be resolved to fully support computer-supported collaborative work, the quality of the information being shared is one of the most significant. The work within the CRC CI shows that solutions to many of these issues are available.

Acknowledgement

The projects described in this chapter were undertaken by the Cooperative Research Centre for Construction Innovation (www.construction-innovation.info). A substantial number of the partners and their personnel participated and their contribution to the overall success of the projects is acknowledged.

Table 3.2 Collaboration issues and the method of resolution

Collaboration issue	Method of resolution
Quantity, quality and level of detail of available information gradually increases throughout the project.	Design View: model that supports all plug-ins. To Do and Problems panes.
Agreed format for information exchange.	Use IFC between external software and CRC CI software. Single schema internally.
Multiple views of project information.	Plug-in architecture around shared representation of the building. Multiple ways of looking at data. Visualisation as a shared function.
Mapping between models and representations. Expanding scope of models through mappings.	Express-X and OCL.
Natural language translation. Clarity of terms.	Prototype link to BARBI server. IFD.
Keeping related information accessible from one interface.	Design Check – encode entire code and results in one place.
Need to be able to restrict (filter) the information being considered at any one time.	Allow users to specify which objects, trades, etc. are of interest at a particular time.

References

Eclipse (2007a). 'Eclipse Modeling Framework', http://www.eclipse.org/modeling/emf/

Eclipse (2007b). 'Eclipse Rich Client Platform', http://wiki.eclipse.org/index.php/Rich_Client_Platform

Eclipse Modeling Framework Project (2008). 'Eclipse Modeling Framework Project (EMF)', http://www.eclipse.org/modeling/emf/ (accessed September 2008).

Eclipse.org (2008). Eclipse.org home, http://www.eclipse.org/

Haymaker, J., Kunz, J., Suter, B. and Fischer, M. (2004). 'Perspectors: Composable, Reusable Reasoning Modules to Construct an Engineering View from Other Engineering Views', *Engineering Informatics*, 18(1), January, pp. 49–67.

IAI (2006). IFC 2× Edition 3, http://www.iai-international.org/Model/R2x3_final/index.htm

OMG (2006). 'Object Constraint Language Specification, version 2.0', http://www.omg.org/technology/documents/formal/ocl.htm (accessed September 2008).

Parlour, R. P. (1997). *Building Services: Engineering for Architects – A Guide to Integrated Design*, 3rd edn. Pymble, NSW: Integral Publishing.

van Leeuwen, J. P. and van der Zee, A. (2003). A Distributed Object Model for CSCW in the Construction Industry, in G. Maas and F. van Gassel (eds), *Proceedings of the International Symposium on Automation and Robotics in Construction (ISARC)*, pp. 221–228. Eindhoven, Netherlands, 21–25 September 2003. Eindhoven: Eindhoven University of Technology.

4 Process modelling for planning, managing and control of collaborative design

Andrew Baldwin, Simon Austin and Paul Waskett

Introduction

Process modelling has become an established tool whereby the information requirements of the design team may be identified, mapped, modelled and then reviewed to ensure that the final design and the design process take cognisance of all the stakeholders involved within the project. Over the last decade several techniques and products based on this approach have emerged. The ADePT technique developed from initial research at Loughborough University has been used for the management of the design of a range of infrastructure projects and engineering products. The methodology that has subsequently evolved has helped to: ensure the rigorous planning of design; control design deliverables; manage customer expectations; assist with change management; and ensure the development of robust process and control systems.

Over this period the approach described in this chapter has been implemented on over 40 projects in the UK and worldwide, with a total contract value of over £4 billion. Evidence of the overall impact of ADePT implementations has been gathered by ongoing research undertaken by the writers, feedback from individual project teams, and independent assessment. The writers' close association with the initial research that led to the ADePT technique, and the subsequent methodology and its implementation have enabled them to fully monitor and review not only the development of the methodology but also its adoption and use for collaborative design across project teams.

This chapter considers collaborative design to be project-based design involving multi-disciplinary teams. It outlines the ADePT technique, the original method, the subsequently developed methodology, the body of methods that have been produced, and their adoption within a collaborative design environment. Three case studies are introduced to highlight the use of process mapping, the repeatable nature of design, and the control of workflow. Lessons learned for the transfer of innovations are also discussed.

The ADePT technique

The development of the Analytical Design Planning Technique, ADePT, has been widely reported. Full details of the technique may be found in Austin *et al.* (1999, 2000, 2002). The following text taken from Baldwin *et al.* (2007) provides a summary of the technique:

> ADePT enables the planning of building design to be approached in a more systematic manner through the use of process modelling to produce a model of the information required, analysis of the models by a technique known as the Dependency Structure Matrix, and the production of design programmes. It provides a way to understand the entire design process by taking a systems view to design. The technique improves the efficiency of the design process by reducing the level of iteration in design tasks, providing an understanding of the effects of change and reducing abortive work. It enables the constraints of earlier design and subsequent construction processes to be managed.
>
> The technique may be viewed as a four stage technique. The first stage involves the production of a model of the design process which identifies the design tasks involved and the information requirements for each of these tasks. (To assist with this task a generic model of the information required at the detailed design stage of a building design comprising some 106 tasks and 104 information flows is available.) The second stage transfers the data into a matrix form, (the Dependency Structure Matrix, DSM), which is used to identify loops within the iteration process. The third stage is the re-arrangement of the task order to break down the iteration block producing an optimised DSM. This enables the programme for the design of the building to be revised based on the optimised design process. The fourth stage enables the output from the DSM matrix to be input into a conventional project planning software package.

Figure 4.1 shows an example from the model. Figure 4.2 shows the four stages of the Analytical Design Planning Technique in diagrammatic form.

From technique to methodology

ADePT has been used across a range of projects and disciplines under different forms of procurement. The 'generic model' developed to provide the basis for building construction has also formed the basis for modelling the information requirements of other civil engineering and construction works. The predominance of public–private partnership programmes for the National Health sector of government has led to a range of successful hospital and healthcare commissions. Similarly the methodology has been adopted by those responsible for a number of new schools and Ministry of Defence (MOD) projects. All these projects have been characterised by their

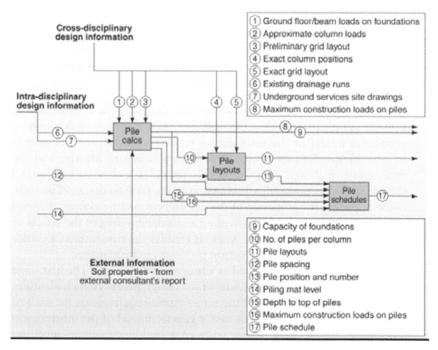

Figure 4.1 An example of the design tasks and information requirements within an information model.

Source: Baldwin *et al.* (2007).

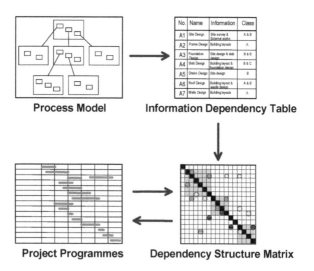

Figure 4.2 The four stages of the analytical design planning technique in diagrammatic form.

Source: Baldwin *et al.* (2007).

multi-disciplinary nature and the need to identify, understand and manage the interfaces between design disciplines and management processes. Experience gained in using the technique has led to a broader methodology for design management.

Use of the technique in a commercial environment highlighted the need for improved management of the design deliverables as 'packages' of work. In addition to the constituent parts of the ADePT technique shown in Figure 4.2, additional tools and techniques based upon spreadsheet methods have been developed to enable the monitoring of this 'workflow', a process integral to the production of design deliverables. The resulting methodology is shown in Figure 4.3. This includes: defining the design process; optimising the design process; producing the project and departmental schedules; and performance measuring and reporting.

Adept Management, a 'spin-off' company formed in 2002 to provide design management consultancy and market the software and associated software products, has developed templates of design activities for use on typical building design projects. These templates contain all the necessary logic and information flows within the design process and may be quickly amended to suit the specific requirements of the project and the procurement activities included to produce coordinated design and procurement programmes.

Using the methodology has highlighted not only the need for managing the design process at the early stages of the design but the need to continually monitor the design process through to the completion of the design. The importance of planning the design process has to be supported by the management of the design process. The ADePT methodology provides a level of detail previously unavailable to senior management, a basis to understand a range of different design solutions and detailed analysis of the information needs of the design and construction teams. Whilst those involved in the production of design deliverables may recognise the benefits of the technique this does not mean that these organisations readily change existing work patterns and fully commit to providing updated production information. An inability to contribute fully may be due to resource availability, contractual restrictions and/or a reluctance to provide information because of commercial sensitivity. Collaborative tools and techniques do not automatically overcome existing business culture.

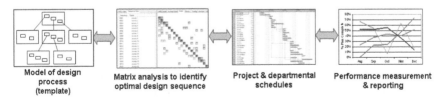

| Model of design process (template) | Matrix analysis to identify optimal design sequence | Project & departmental schedules | Performance measurement & reporting |

Figure 4.3 The further developed ADePT methodology.

ADePT has been particularly successful when linked with the approach to planning identified as 'Last Planner' (see Ballard and Howell 1994; Ballard 2000). Planning that adopts the Last Planner philosophy 'follows a production management philosophy that includes reliability scheduling and controlling design activities' (Choo *et al.* 2004). Combined, Last Planner and ADePT have been termed 'DePlan'. This methodology

> helps planners to generate quality plans, that is, plans that express what is ready for execution; by sequencing activities in the right order; by identifying informational and resource requirements ahead of design execution; by sequencing activities in the right order; by identifying informational and resource requirements ahead of design execution; and by scheduling only those activities that have met these requirements.
>
> (Choo *et al.* 2004)

It focuses on what can be achieved with the resources currently available whilst highlighting the design tasks that are unable to be commenced. The overall ADePT approach, incorporating DePlan, is summarised in Table 4.1.

This way of working and its benefits may be summarised as follows:

- It identifies and removes turbulence from the project process.
- It provides greater certainty of design coordination.
- It offers the ability to better prioritise design work.
- It integrates subcontractor design with consultant design in an effective way.
- It makes management of design change more effective than is typically the case.
- It improves collaboration between design team members.
- It focuses the team through workflow control to task completion.

To achieve these benefits it has been found essential to adopt a facilitated approach to planning whereby a facilitator

> defines the high level structure of a design plan; involves the design team members at appropriate times in defining the design scope and identifies issues around the interfaces between design, procurement and construction – and enables a consistent and meaningful programme to be produced.
>
> (Choo *et al.* 2004)

One party has to take responsibility for the control and production of the design information. This role of design manager, responsible for the modelling, analysis and subsequent monitoring of the design deliverables,

Table 4.1 The ADePT approach incorporating DePlan (Choo et al., 2004)

Step	Description of ADePT work	Features of approach
1	Modify design templates to suit requirements of the project, incorporate the contractor's procurement activities and start on site milestones.	All design tasks fully integrated with similar levels of detail. All undertaken within same planning environment. All logic inherent across disciplines and with procurement included.
2	Undertake an initial streamlining of the design and procurement process by considering initial design compromises to unlock the design process.	Design activities coordinated across the disciplines and procurement. Areas of collaboration needed to deliver an efficient design highlighted (interdependent blocks). Initial design risks and design assumptions produced to deliver initial design programme.
3	Export the initial ADePT model into planning tool, add and level resources and impose procurement and start on site deadlines.	Resource design activities to give a levelled programme, coherent for design disciplines and project as a whole. Highlights pinch points in the project process. Release of design information streamlined to suit construction sequence.
4	Consider effect of undertaking design sub-optimally and resolve pinch points.	Procurement dates for work packages identified. Change to suit design or procurement programme, whichever takes priority. Realistic release of information from the design team produced.
5	Produce a single, coordinated and realistic multi-disciplinary design and procurement programme.	Potential to reduce the number of work packages issued identified. Release of design information for construction streamlined into natural clusters.
6	Issue work plan to design team covering short-term look-ahead period.	Design team can focus on a 'to do' list rather than programme. Constrained activities identified and action plans put in place. Scheduled activities cover only those free from constraints.
7	Capture progress, update programme and report performance.	Programme updated regularly. Design team performance against programme reported. Impacts of delay between team members highlighted.

may be allocated to one of the organisations involved in the project or may be undertaken by an outside consultant.

Case study examples

Case study 1 – process mapping

The Highways Agency introduced their early contractor involvement (ECI) form of contract to gain benefit from contractors' input at the early stages of design on new highway schemes. This early input would ensure the design solution could be built safely and efficiently, was maintainable and was sustainable. Having to introduce their project managers into the design teams at an earlier stage meant that the roles of the contractor's project manager and the consultant's design manager, and the interfaces between these roles, needed to be defined.

Adept Management worked with a major civil engineering contractor and their five design partners to develop process maps of the early stages of the highway design process and of the corresponding management processes. The design process maps covered six key disciplines: traffic and economic analysis; structures design; highways design; geotechnical design; environmental analysis and design; and statutory procedures. It was clear from the maps produced that the highways design process was the central element of the overall design process, since a high proportion of the information exchanges between design team members passed through this design discipline. The design process maps were developed with input from all five design consultancies. There was a very high level of commonality of approach between the companies, with only minor differences around terminology and no fundamental variations. This reinforces one of the basic premises behind the ADePT approach: that generic templates can be produced and implemented on a range of projects because, whilst the product may vary, the design process is largely generic.

The management process maps covered 18 aspects of the project management, such as risk management, value management and team communication. All cross-disciplinary information dependencies were identified, whether they were across the design disciplines or between design and project management. Of course, these were based on the team's current understanding of projects rather than any experience under the ECI contract. So, as expected, there were interfaces between designers and the contractor's project manager where the direction and timing of information exchange were unclear. However, the integrated set of process maps provided valuable insights to the team over where the project manager should intervene in the design process and the division of responsibilities between project manager and consultant's design manager.

In this study, as is often the case where process mapping is seen as a useful tool, the motivation is to introduce a new or improved process. However, the new process is something of an unknown and so an existing process is what is mapped, thus allowing waste to be identified and removed and the new process to emerge. Adept Management used dependency structure matrix (DSM) analysis to identify interdependence in the integrated highway design process which had been mapped. Key decision points could then be identified, and the timing of information exchange between designers and the project manager could be clarified. This analysis was, in effect, the first step in redesigning the process to suit the ECI contract.

The civil engineering contractor and its design partners now have the process map available as a template, thus enabling robust integrated action plans to be put in place on future ECI contracts.

Case study 2 – the repeatable nature of design

When the MOD commissioned the redevelopment of one of its largest garrisons, the challenge facing the design team was significant. The site was to incorporate around 130 buildings in total and, whilst the overall project timescales did not present a major problem, some of the design deadlines were extremely tight, being dictated by the MOD's design check procedures and timescales. In addition, with so much design information to be produced, managing the resource requirement on the design team was a critical element.

The consultant was developing a range of standard solutions to be rolled out across the buildings. With such a wide (and fast) roll-out it was imperative that the design solutions were fully coordinated first time, as any problem would need to be dealt with as many times as the design was rolled out.

The multi-disciplinary design consultant wanted to put in place design programmes for each of the buildings. Rather than plan each building's design independently, Adept Management worked with the consultant to identify a small number of 'generic' building types, such as training buildings, accommodation, office space and so on. Then Adept Management's generic building design templates were used to develop templates for each building type, which could be tailored quickly to suit each individual building.

The templates produced were used to develop design programmes using the DSM stage of the ADePT approach. This then gave the consultant a suite of design programmes, each incorporating activities associated with key coordination points and integrated with the MOD's design check and sign-off processes. Developing this overall suite so that the consultant's resources could be moved seamlessly across buildings was a major challenge. One of the difficult aspects of this was in understanding the timescales needed for the design of each building where later buildings were, in some cases, largely a roll-out of a previous design, with only the building's interface with the ground requiring any real new design information. The temptation was to slash timescales, but any error in estimating timescales could have led to a large delay as a proportion of the overall time allowed for the design. The consultant monitored the time required for design as solutions were rolled out across the project and could very quickly gauge what reduction in time it could expect on future buildings.

This case study shows the highly repeatable nature of the design process, not just because on this project design solutions were being rolled out but because a small number of design processes could be applied to a large number of buildings. The overall project was planned with effort that did not extend significantly beyond that expected for a single large building.

Case study 3 – workflow control

Adept Management was commissioned to develop a suite of design programmes on a major healthcare development in the north of England, comprising two new hospitals. The buildings were commissioned by the government under a Private Finance Initiative (PFI) contract, and so the sign-off and approval processes involved in that form of contract had to be built into the design process models. The result was that the highly detailed programmes highlighted not only the cross-disciplinary coordination issues but also the contractor and client cost checks, review meetings and approval points. Ultimately the programmes featured contractor and client activities as heavily as some of the design team consultants, showing the importance of their roles in an integrated design process.

The fourth stage of ADePT, control and monitoring of workflow, was implemented on the project. Periodically, typically each month, a schedule of activities was distributed to each of the design team members, including contractor and client. This highlighted the activities which were due for completion in that period, those which were due to be progressed, and those due in the following period, so priorities could be assigned. The team members reported back at the end of each period the progress against each activity, any reasons why activities due for completion had not been completed, and any known constraints affecting forthcoming activities. Based on this feedback, the programmes could be updated and any delays or constraints could be actioned.

The progress across the team could give an overall indication of progress, identified by combining all progress on the programme. However, the management team on the project were also interested in seeing the performance of team members individually. This was measured as the ratio of activities due for completion in a given period to the activities which were actually completed, called percentage planned complete (PPC) in the Last Planner technique.

This measure was used since it is only upon the completion of a design activity that all of its outputs can be said to be fully coordinated and complete. The measure focused the team upon fully completing activities, since a report that all activities were 90 per cent complete scored a PPC of zero. So the scenario where an activity's progress developed over time by 0, 50, 80, 85, 90 and 95 per cent could be replaced by one where progress developed by 0, 50 and 100 per cent. Feedback collected from the project's design director confirmed that the approach had the desired effect, focusing the team on completion of tasks.

Lessons learned

The ADePT methodology has become a proven technique for improving the management of the design process. The methodology has shown how process modelling supplemented by use of the DSM technique, when linked to traditional project planning software, can provide a comprehensive toolkit for design management. Presentation of the output from ADePT to traditional project planning software enables the impact of design decisions on the design and construction process to be fully evaluated. The output of ADePT is best linked to planning software applications, for example MS Project, Primavera, Power Project and so on, as it is then possible to produce the detailed level of analysis required for effective planning. The adoption of the Last Planner philosophy has been found particularly appropriate in the effective management of the design process.

Experience of using the technique in a business environment has highlighted a number of changes in methods of working from those developed in the pilot testing during the research. The use of the Information Dependency Table (see Figure 4.2), found to be a good working tool to record experts' decisions in a research environment, has not proved efficient in a commercial environment, members of the design team preferring to move as swiftly as possible to the project planning software environment to review the results of the analysis. The generic building model developed to form the process model required for all individual projects is not always the most suitable basis for new model developments. In many cases the design team prefer to use models already produced for previous projects, as this results in an initial model for use by the design team.

The primary focus of ADePT is the management of the information flows between designers, not just design deliverables. However, the production of design deliverables to agreed time schedules remains the primary concern of design managers. The workflow tools and techniques developed to assist the design manager in this task have been important in supplementing the basic technique which utilises DSM to investigate and optimise design solutions. The use of ADePT is important in enabling design managers to fully understand the implications of incomplete information or assumed information. From the feedback obtained from the projects where the software has been used it has been necessary to continually develop the software to meet the changing needs of users. At the time of writing, two commercial software tools are in development to enable the planning and control aspects of ADePT, named ADePT Design Builder and ADePT Design Manager respectively.

Conclusions

In 1998 the *Rethinking Construction* report highlighted that the separation of design from the other phases of a project was a fundamental weakness in the construction industry (Egan 1998). During the subsequent period the continued utilisation of PFI and design and build as preferred forms of project finance and procurement has highlighted the need for improved management of the design process. Despite this focus, the problem of late delivery of design information and its subsequent impact on construction schedules persists. Process modelling can help to overcome these problems. Adoption of the process modelling and related techniques on construction projects has now been shown to provide significant, measurable cost savings.

The introduction of these new tools and techniques is considered by the writers to be imperative for new forms of procurement that bring together teams that are required to embrace the challenges of projects in new ways. The current availability of such systems and the cost of the related technologies is not an obstacle to their adoption. The challenges, as with most systems implementation, are human and organisational. Traditional software systems, particularly project planning systems, remain the preferred communication platform of business organisations because of the familiarity with input requirements, the output produced and the existing ways of working that have been developed around them. The adoption of new techniques on the projects that have adopted process modelling to date has been almost exclusively based on the use of consultants with a record of design management experience. This directly reflects the evolution and introduction of critical path planning methods, the forerunner of current project management software, in the late 1960s and early 1970s. These critical path planning techniques, initially developed in the late 1950s, took some 20 years to evolve, mature and become widely accepted as the pan-industry approach to project monitoring and control. Process modelling techniques for the management of design have now become accepted practice for some construction and engineering organisations. The next decade will show how widely they become adopted across the industry.

Acknowledgements

This chapter was produced with data and material from Adept Management Ltd. For details see http://www.adeptmanagement.com/.

References

Austin, S., Baldwin, A., Li, B. and Waskett, P. (1999). 'Analytical Design Planning Technique: A Model of the Detailed Building Design Process', *Design Studio*, 20(3), April, pp. 279–296.

Austin, S., Baldwin, A., Li, B. and Waskett, P. (2000). 'Analytical Design Planning Technique (ADePT): A Dependency Structure Matrix Tool to Schedule the Building Design Process', *Construction Management and Economics*, 18, pp. 173–182.

Austin, S., Baldwin, A. and Steele, J. (2002). 'Improving Building Design through Integrated Planning and Control', *Engineering Construction and Architectural Management*, 9, pp. 249–258.

Baldwin, A. N., Shen, L. Y., Poon, C. S., Austin, S. A. and Wong, I. (2007). 'Reducing Waste in High Rise Residential Building by Information Modelling at the Design Stage', *Surveying and the Built Environment*, 18(1), June, pp. 51–62.

Ballard, G. (2000). 'The Last Planner System of Production Control', Ph.D. thesis, University of Birmingham, UK.

Ballard, G. and Howell, G. (1994). 'Implementing Lean Construction: Stabilising Workflow', Proceedings of the Second Annual Meeting of the International Group for Lean Construction, Santiago, Chile, available in L. Alarcon (ed.) (1997). *Lean Construction*. Rotterdam, Netherlands: Balkema.

Choo, H. J., Hammond, J., Tommelein, D., Austin, S. and Ballard, G. (2004). 'DePlan: A Tool for Integrated Design Management', *Automation in Construction*, 13, May, pp. 313–326.

Egan, J. (1998). *Rethinking Construction*. London: Department of the Environment, Transport and the Regions.

5 Combining 3D models and simulations to meet the design challenges of the twenty-first century

Souheil Soubra

Abstract

No part of the economic community can escape from the urgent issues related to global warming, carbon footprint and reducing energy consumption. Nevertheless, the building sector is particularly under pressure. Indeed, it is one of the biggest consumers of energy, either directly for lighting and thermal comfort (heating and air conditioning) or indirectly for the production of building materials. It also largely contributes to the massive use of some critical resources (such as energy, water, materials and space) and is responsible for a large portion of greenhouse gas emissions (Ratti *et al.* 2005).

At the same time, the construction sector is expected, more than ever, to provide better living and working conditions: accessible and comfortable for all, safe and secure, durably enjoyable, efficient and flexible to changing demands, available and affordable.

Current business models and working methods have reached their limits, and there is an urgent need for creativity-enhancing tools that support an 'out-of-the-box' approach to design, aiming for:

- environmentally sustainable construction (in a context of limited resources – energy, water, materials and space);
- meeting clients' and citizens' needs in terms of health (from indoor and outdoor exposures), security (against natural and industrial hazards), accessibility and usability for all (including the disabled and elderly), and enhanced life quality in buildings and urban environments.

In that context, the chapter explores the possibilities of using geospatial information as input data to construct 3D models of the built environment. The models are then combined with simulations in order to address sustainable urban development issues within the planning process. Special focus will be given to 1) minimizing energy consumption and 2) simulation of air quality, taking into account meteorological data and traffic conditions.

Finally, it is now commonly agreed that research must not be concerned solely with technology, as social, organisational and human issues also need to be considered in an interdisciplinary manner (Soubra *et al.* 2006).

New working methods need to emerge in order to move away from the current situation where different groups or departments involved in urban planning (e.g. city planning, the legal office, the environment office, the roads department, the green department, etc.) sometimes work on the same project without communicating or, even worse, while hiding crucial information from each other. The chapter will report on how these aspects have been tackled by considering two test cities in Europe.

Introduction

Cities are dynamic living organisms that are constantly evolving, which makes city planning a very complex process involving a large number of actors: planners, developers, communities, environmental groups, local and national governmental agencies, civil security organisations and so on. All of them represent their own interests, the interests of an organisation or the public interest. Handling this complex process is essential, since it has a direct impact on the quality of life of all citizens and the economic, social and governance manifestations of the citizenship.

Currently, the complexity of the city planning process is amplified by the overwhelming number of challenges facing society. Among these, global warming, climate change and the end of low-cost fossil resources are of paramount importance. Thus the quality of our environment is now a major concern for citizens and, as a consequence, their political representatives.

In the meantime and from a technological point of view, society has amassed in recent decades an enormous amount of digital data about the Earth and its people. This tremendous development is very promising, since spatially referenced information is essential for any urban information infrastructure. The ability to seamlessly combine data from different sources and share it between many users and applications is a major issue, and several initiatives (including the Open Geospatial Consortium and IAI IFG) have been launched in order to improve interoperability and the ability to access, process and ultimately understand geospatial information from multiple sources. As these developments progress there emerge opportunities for society as a whole to develop and use intelligent tools that support new, multi-disciplinary, more inclusive planning processes. Such tools would allow citizens, commercial bodies and government employees to explore, in an interactive way, the huge sets of information, thus supporting the user in making sense of it.

Mapping these two dimensions (i.e. new complex interdisciplinary issues along with new technological possibilities) underlines the urgent need and the technical feasibility of innovative design and planning tools that enhance creativity and allow the investigation of 'out-of-the-box' solutions that are not currently pursued owing to their complexity. We consider that this emerging generation of tools will integrate various dimensions, such as linking with geographical information, constructing 3D models of the built environment and using numerical models and simulation tools in order to predict the behaviour and impact of proposed design and planning options. These tools can allow the tackling of the highly complex and interdisciplinary challenges facing society (e.g. reduction of energy consumption, taking into account various physical forms of the built environment and various estimates of climate change) through new possibilities.

The twenty-first-century context: dwindling fossil resources and global warming

The Earth's global mean climate is determined by incoming energy from the Sun and by the properties of the Earth and its atmosphere, namely the reflection, absorption and emission of energy within the atmosphere and at the surface. Changes have occurred in several aspects of the atmosphere and surface that alter the global energy budget of the Earth and can therefore cause the climate to change (IPCC 2007).

Among these are increases in greenhouse gas concentrations, which act primarily to increase the atmospheric absorption of outgoing radiation, and increases in aerosols (microscopic airborne particles or droplets), which act to reflect and absorb incoming solar radiation and change cloud radiative properties. The dominant factor in the radiative forcing of climate in the industrial era is the increasing concentration of various greenhouse gases in the atmosphere. Several of the major greenhouse gases occur naturally, but increases in their atmospheric concentrations over the last 250 years are due largely to human activities. As an example, 300 billion tons of CO_2 have been emitted by human activities since 1750. It is estimated that about two-thirds of anthropogenic CO_2 emissions have come from fossil fuel burning and about one-third from land use change. About 45 per cent of this CO_2 has remained in the atmosphere, while about 30 per cent has been taken up by the oceans, and the remainder has been taken up by the terrestrial biosphere. About half of a CO_2 pulse to the atmosphere is removed over a time scale of 30 years, a further 30 per cent is removed within a few centuries, and the remaining 20 per cent will typically stay in the atmosphere for many thousands of years.

The impact on the climate is still subject to discussion. Nevertheless IPCC concludes that:

- it is *extremely unlikely* (less than 5 per cent) that the global pattern of warming observed during the past half-century can be explained without external forcing;
- it is *very likely* that anthropogenic greenhouse gas increases caused most of the observed increase in global average temperatures since the mid-twentieth century.

Finally, projections of future changes in climate are also hazardous, even though confidence in models has increased owing to improvements in

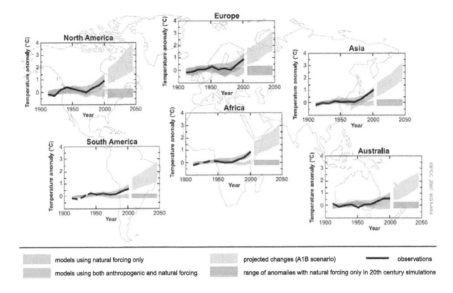

Figure 5.1 Continental surface temperature anomalies: observations and projections.

the simulation of several aspects of the climate. An increase of several degrees (probably between 2°C and 5°C) by 2100 is generally accepted (Figure 5.1).

The scale of the future changes in climate is also largely indeterminate, since it depends on the capacity of our society to react and reduce greenhouse gas emissions. The reduction aimed for in the Kyoto protocol is of 5.2 per cent for signing parties by 2008–2012. The European Council decided the '3×20' objectives by 2020 independently of any international agreement:

- 20 per cent reduction of greenhouse gases;
- 20 per cent reduction of energy consumption;
- 20 per cent use of renewable energies.

This roadmap has, among other actions, been adopted recently by the French 'Grenelle de l'Environnement' conference.

The construction sector: an industrial champion of greenhouse gas emissions?

Basically, each and every act of our everyday life generates greenhouse gases. It's obvious for some of our actions (driving, flying, etc.) but may be less so for others (eating, because of the plastic of the packaging or the fertilizers used, etc.). Aiming for the '3×20' reduction strategy implies giving priority to greenhouse gas emissions in our day-to-day actions.

Recent studies show that greenhouse gas emissions in France were overall stable between 1990 and 2004. Unfortunately, behind this average figure lie important differences: several sectors reduced their emissions (industry, energy, agriculture, etc.), while others significantly increased theirs. Among these were transport and, very close behind, the construction sector.

In France, the construction sector generates up to 25 per cent of the greenhouse gases and uses 19 per cent of the overall energy and even up to 46 per cent of the overall energy if we include electricity and heating used in buildings. Any serious strategy aiming to reduce greenhouse gas emissions and make energy savings will have to include the construction sector.

Combining building design and urban planning

Addressing complex issues such as global warming, the carbon footprint and reducing energy consumption cannot be done without taking into account the interactions in both directions between cities and buildings (i.e. how the environment impacts the buildings and how the buildings modify the environment). These interactions are becoming more and more complex. Obviously, cities, buildings and their various elements need to be integrated as complex systems of material, energy and information flows, but the lack of adapted models and tools doesn't allow design to move towards this needed holistic approach.

Ken Yeang (1996) considers that bioclimatology, in architectural terms, is the relation between the form of a structure and its environmental performance in relation to its external climate. Although such an approach has higher start-up costs, it produces lower lifecycle energy costs, as well as providing a healthier and more human environment within the structure. Other issues he considers vital to bioclimatic consideration are those of place-making, preserving vistas, creating public realms, civic zones, physical and conceptual linkages, and the proper massing of built forms.

Despite that, assessing both the environmental performance of buildings and the impact of a building on its site are still largely based on expert guessing supported by simple, disconnected, mono-disciplinary models that cannot support the holistic approach needed.

Environmentally sustainable design should be considered as a holistic process where architects, urban planners and engineers must share a solid knowledge of local conditions, existing resources and the possibilities of using renewable forms of energy and materials. Tools supporting this holistic approach still need to emerge. They will obviously allow linking with geographical information and GIS and combine these with CAD models and simulation packages.

Integrating geographical information

This vision relies on the seamless integration of multiple sources of digital data. As these data have been amassed by many different organisations over several decades a preferred solution is to adopt open data standards for handling data, towards which legacy systems can then migrate. Fortunately several industry-led consortia have been promoting standards relevant to urban planning. The mission of the Open Geospatial Consortium (OGC), for example, is to deliver interface specifications for geographically referenced data. To this end it has developed several specifications to describe, encode and transport information pertaining to spatial entities. In contrast to Industry Foundation Classes, a common building information model developed for the construction industry by the International Alliance for Interoperability, the OGC specifications apply to generic spatial entities such as points, lines and areas and need to be contextualised before they can be used for a particular application.

As an example, an application schema for city planning was developed by CSTB according to OGC Geography Markup Language (GML) specifications. GML is an XML-based mark-up language that is used to describe geographical phenomena and encode information about them. These features have geometry and non-geometry properties, but feature types, such as roads, houses or municipal boundaries, are not defined in GML itself. They are defined in application schemas.

The schema separated common elements from elements specific to an application, such as those that describe certain parameters of the environmental simulations and visualisations. In this way a Planning.gml file was generated by mapping the .tab data files of two test cities to the required specific and common elements of the .xsd schema using an XSLT style sheet transformation. The common elements included a FeatureCollection and AddressType and inherited the elements of CityGML (http:// www.citygml.org), a GML3 schema for three-dimensional city models (Harrison 2005). The specific elements included building, noise barriers, road centreline network and 2.5D building.

GML-encoded information can be easily transported over the Internet. It can be used to communicate descriptions of feature types or transport actual data for a particular feature. A parser was developed allowing GML files to upload through a MapServer implementation, with the data delivered as a Web Feature Service to a Mozilla Firefox browser or displayed in the visualisation engine of the environmental simulators (Figure 5.2).

Simulation tools

Being able to access information about environmental dimensions of building materials and components (e.g. carbon footprint, energy performance, etc.) and combine it with simulation packages allowing prediction of the

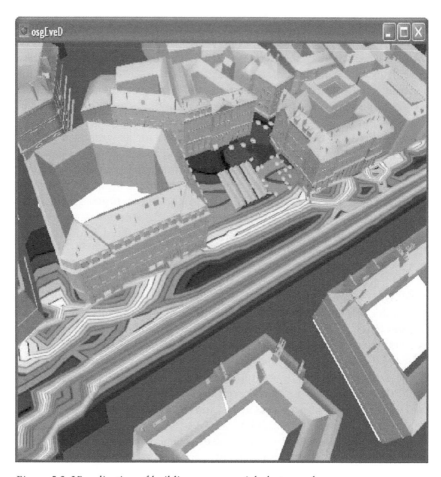

Figure 5.2 Visualisation of buildings on an aerial photograph.

effects of using different materials or components in various conditions will produce significant advantages to the designer from an environmental point of view. The design can take into account the knowledge about the environmental impact of the building over its lifecycle. Many factors can be considered, including the cost implication of using smart materials, solar gains, and the many trade-offs associated with the use of potentially lower-carbon-footprint materials and their energy performance.

Application: climate change, thermal comfort and energy use

There is little need to justify the drive to minimise energy used in buildings, even though motivation seems to have shifted from reducing cost to saving scarce resources and, recently, to minimizing the production of carbon dioxide.

For a given set of climatic conditions, building energy performance is currently understood as dependent upon:

- urban geometry;
- building design;
- systems efficiency;
- occupant behaviour.

Complexity comes from the fact that these four points are addressed by four different stakeholders, namely:

- urban planners;
- architects;
- system engineers;
- occupants.

Obviously, any serious work aiming to reduce energy consumption needs to support collaboration and the sharing of in-depth knowledge of all functional, technical and design relationships between these stakeholders.

According to Baker and Steemers (2000), building design accounts for a ×2.5 factor in the variation of energy consumption. System efficiency and occupant behaviour account for a ×2 factor each. The cumulative effect of these factors lead to a total variance of ×10. Nevertheless, it is observed that variance in energy consumption of buildings with similar functions can be as high as ×20. Urban geometry is assumed to be the missing ×2 factor, but research has not yet been carried out on this topic to validate this assumption. Indeed, even if the interaction between the building's energy balance and the surrounding through the envelope is well known, the interaction between the building and the urban geometry is underestimated and usually neglected, probably owing to the complexity of the environmental processes involved. The outdoor temperature, wind speed and solar radiation to which an individual building is exposed is not the regional climate but the local microclimate as modified by the physical form of the city mainly in the neighbourhood where the building is located (Giovani 1989).

The specific case of thermal comfort under warmer summer conditions will be a key challenge for designers in the twenty-first century. Solutions will move toward low-energy buildings that are free running for some parts of the summer (being either entirely naturally ventilated or mixed mode, where mechanical cooling is used only when thought to be essential). Because internal conditions in these buildings will vary from day to day dependent on the climate conditions, understanding users' behaviour and reactions to adapt to the environment will be of paramount importance. Therefore we assume that, for low-energy designs that are, by definition, more sensitive to climate change and user behaviour, the ×2 factors for both of these aspects will probably be increased in a meaningful way.

a. Wasteland between Jardines del Viejo Cauce del Turia and port, Valencia.

b. Ciudad de las Artes y las Ciencias in the Jardines del Viejo Cauce del Turia, Valencia.

c. Container port area planned for redevelopment, Aarhus.

d. Opening up the city centre by redeveloping the river corridor, Aarhus.
Figure 5.3 Urban (re)development in Aarhus and Valencia.

Examples of Aarhus and Valencia

As it is now commonly agreed that research must not be concerned solely with technology, experimentation has been launched with two test cities in Europe (Aarhus in Denmark and Valencia in Spain) in order to assess the impact of new working methods on organisational and human issues.

Consultations were held with urban planning professionals in Aarhus and Valencia to understand their planning goals and priorities and identify the planning process and information flow. It was observed that there exist differences in the legal systems, instruments and documentation but that in practice there was considerable similarity in the business process. Typical scenarios were discussed and documented. The visual appearance and environmental consequences of planned developments were important to both cities. Scenarios were validated by checking their feasibility against data, procedures, policies and strategies, organisational and political structures, and the technological capacity of the cities.

Conclusion

Pressure to study energy conservation is increasing rapidly owing to soaring oil prices and concerns about climate modifications. In the construction sector, the first step to reduce energy consumption is to simulate the behaviour of buildings. Many energy models have been developed over the years, but these models tend to represent the buildings as separate entities and neglect the importance of the interactions between the building and the urban scale. One of the reasons for this lack is the difficulty of modelling complex urban geometry. The integration of geographical information in the building design process seems therefore inevitable. From a technical point of view, proceeding to this integration through open data standards (e.g. OGC OpenGIS) is a relevant approach to address multiple data sources. An implementation using OGC GML showed the technical feasibility of this integration and the ability afterwards to address technical simulations.

In parallel, a study on soft issues showed that despite apparent differences in the planning regime of northern and southern European countries there is sufficient similarity in their business processes to make the development of reusable components worthwhile. This research has demonstrated that OGC specifications such as GML and WFS can be adopted as part of an e-planning business service and used in the communication between this service and other spatially enabled services in the service-oriented architecture. The success of similar frameworks will not, however, be determined on technical grounds. The really big challenges are to develop a viable and sustainable business model and to establish a culture of openness and cooperation in the corridors of city halls.

Finally, the construction sector is facing an overwhelming number of complex issues with a high interdisciplinary aspect. These issues stress the need for creativity-enhancing design methods and tools that allow the investigation of 'out-of-the-box' solutions that are not currently pursued owing to their complexity. More specifically, research objectives need to include the following:

- Buildings and their various elements need to be integrated as complex systems of material, energy and information flows. Models linking city planning issues and sustainability-related issues on both the urban and the building levels need to be developed.
- Various disciplines (technical performances of the envelope, comfort conditions, health and safety, cost analysis, etc.) need to be integrated in an interdisciplinary and holistic approach supporting a 'global optimum' search that allows designers to confront options in order to identify and handle conflicts.
- Consultation with clients and citizens is a definite must in order to put the users at the centre of the process and to envisage a sound operation of the building. New design paradigms need to emerge where client brief and design are conducted in parallel and influence each other as they progress.

References

Baker, N. and Steemers, K. (2000). *Energy and Environment in Architecture.* London: Spon.

Giovani, B. (1989). *Urban Design in Different Climates*, WMO/TD no. 346. Geneva: World Meteorological Organization.

Harrison, J. (2005). 'Partners Pioneer GML Deployment for NSDI', *Directions Magazine*, 19 September.

Intergovernmental Panel on Climate Change (IPCC) (2007). *Climate Change 2007: Synthesis Report.* Geneva: IPCC.

Ratti, C., Baker, N. and Steemers, K. (2005). 'Energy Consumption and Urban Textures', *Energy and Buildings*, 37(7), pp. 762–776. Elsevier.

Soubra, S., Marache, M. and Trodd, N. (2006). *Virtual Planning through the Use of Interoperable Environmental Simulations and OpenGISfi Geography Mark-Up Language.* Dublin: eChallenges.

Yeang, K. (1996). *The Skyscraper Bioclimatically Considered: A Design Primer.* London: Wiley–Academy.

6 Digital design collaboration

From BIM to BKM – enhancing human creativity

Rivka Oxman

Abstract

Computational enhancement of human collaboration in digital design demands a shift from information-based technologies to knowledge-based technologies. Replacing the concept of the 'building information model' (BIM) with the concept of the 'building knowledge model' (BKM) may support such a shift in supporting human collaboration in digital design. In order to illustrate the potential of this approach, we present two cognitive models of experiential design knowledge. These two specific cognitive models of design collaboration are reviewed and their relevance to BKM are presented and discussed. The first supports creative collaboration presenting the phenomenon of schema emergence in the human mind and in digital design media. This approach refers to the emergence of typological knowledge and topological structures. The second model, termed issue–concept–form (ICF), is a cognitive model that captures the design rationale and expertise. It provides explicit contextual knowledge of prior design precedents to support collaborative work. It is proposed that these types of knowledge should be captured in a future BKM technology to support human creative collaboration in digital design.

1 From BIM to BKM – creative design and human collaboration

1.1 Introduction

Human collaboration is based on the ability to explicate and share knowledge. Despite the fact that BIM technology has contributed to information exchange and communication in collaborative work, it cannot yet be considered as a medium that supports design innovation and creativity. BIM technologies do not yet provide a knowledge medium which can support creative collaboration among team members employing digital media acknowledging the ways that human designers behave and interact.

Research in the field of BIM needs to go beyond questions of project documentation, coordination and integration in order to encourage collaboration between members of the design team. Human capabilities such as design innovation and creativity must be addressed and supported in digital design in the same way as efficiency and correctness are today.

The aim of this chapter is to introduce certain types of design knowledge that are employed in collaborative processes employing digital media. The ability of innovative technologies to capture this type of knowledge may contribute to the shift that will allow us to move from BIM to BKM. We consider BKM as models of high-level knowledge in design collaboration that can potentially enable the correct and timely exploitation of information components of BIM during the various stages and functions of the BIM. Furthermore, BKM, being high-level knowledge, can further enhance the capability of advanced BIM to contribute to the innovation and creativity of the design–engineering–construction and management project team.

1.2 BIM – a brief introduction

Current research in, and applications of BIM have contributed to the development of certain digital properties that assist and improve the design team's performance in collaborative and integrated activities. BIM technology has been proved to enhance the communication and sharing of building information. The BIM used by the AEC professions generally covers geometry, spatial relationships, geographic information, quantities and properties of building components for various collaborators (Schevers *et al.* 2007). Each member of the team can add his/her discipline-specific knowledge referring to a single model. Structural engineers, mechanical engineers, electrical engineers and so on can inform each other, the contractor and the owners regarding specific detailed aspects of the building model.

Currently, the main benefit of BIM is its ability to decrease errors made by design and construction teams by employing the mechanisms of conflict detection through visualization techniques, referring to relevant parts in relation to the whole building model. As BIM becomes more capable of handling more building information, this can help to save costs through reduction in time and errors. One of the leading projects is the success story of the Swire Properties Hong Kong East Tower. The Swire Properties Limited One Island East Tower design and construction project in Hong Kong is one of the most substantial BIM implementations ever undertaken. Gehry Technologies (GT) were employed in an innovative implementation of new BIM technologies and working methods to help to realize saving in the cost of construction, by enhancing efficiency and reducing waste across the entire process (Riese 2007). One of the major contributors to this project was the development of a 4D simulation tool developed by Heng Li and the Virtual Prototyping Laboratory at the Hong Kong Polytechnic University, in cooperation with GT (Huang *et al.* 2006). Incorporating Gammon's construction expertise with the power of process visualization contributed to a unique collaboration process. Currently these kinds of preventive and efficiency-producing characteristics of team interaction across the life cycle are considered the main attributes of BIM.

This chapter suggests that a shift from the concept of BIM to the concept of BKM can potentially provide positive aspects in the support of human multi-disciplinary collaboration in the multiple stages of design. In the following section we present and discuss issues for achieving computational environments that provide knowledge and that can potentially recognize, generate and accommodate the informational requirements of a design according to the dynamics of a multi-disciplinary, multi-stage perspective.

2 Design knowledge and creative collaboration in digital design media

Competence in design collaboration appears to be measured by knowing which specific kind of knowledge and information to apply in a particular situation, how to apply it when needed, and how to accommodate multi-disciplinary differences. In considering the definition of design knowledge as relevant to the questions at hand and to future integration in BKM, the following issues are raised:

- What types of knowledge do designers use, and how is it integrated and communicated?
- How do team members exchange knowledge and expertise in various stages of collaboration?
- How can this knowledge be captured computationally?

Design knowledge may be defined as the knowledge of how to generalize, organize, apply and communicate knowledge gained experientially. Design knowledge may also be a significant class of knowledge that helps designers to organize and exploit factual knowledge or information. It is this capability of organizing, interpreting and communicating information that is particularly relevant to the advancement of BIM. In this sense, the BKM is a meta-model of the BIM that contains higher-level knowledge that enhances the application and communication of the information of the model relative to the particular discipline and the particular stage. Cognitive models suggest how experiential knowledge can be formulated and how it might be potentially applied to assist design collaboration.

In the following sections we present two relevant examples of the formulation of cognitive models of design knowledge that can be applied in collaboration. The first is based on high-level knowledge of design typologies, and the second is based on an important class of experiential knowledge – the knowledge of known precedents. These two models are revisited and presented, and their relevance to and potential application in BIM technology are presented and discussed. Such experiential and typological knowledge can potentially enhance the creative capability as well as the preventative functions of current BIM.

3 Creative collaboration in digital design

The reinterpretation and restructuring of design representations are a fundamental property of visual reasoning in creative collaboration. This reasoning process is generally facilitated by the interaction between design collaborators by the visual representations of the design. The process of recognizing new emergent properties within existing representations characterizes what is termed *emergence* (Oxman 1999). Our theoretical assumption is that generic knowledge is one of the forms of knowledge that experienced designers know how to employ in the process of reinterpreting, reformulating and modifying designs. The process is common to all members of the team. While particularly important in the early phases of architectural and engineering design, the logic of visual reasoning and emergence is relevant to all stages.

In this chapter we report on an experimental work towards the support of visual emergence in creative collaborative design (Oxman 1999). We have investigated how emergence can be utilized to support two designers or teams that share and participate in a *shared representational environment*. Within this representational environment knowledge of design may enhance the possibility of the emergence of a new schema. In the following section conceptual issues in visual emergence in collaboration are discussed.

3.1 Schema emergence (morphogenesis)

The emergence of a new schema is a fundamental cognitive capability of creativity in the human designer. A paradox of creative design is how the human designer can discover a new schema while working with the existing context of an existing schema. A recognition of these phenomena of innovation and creativity can be seen in the transformation process in which specific prototypes (of chairs) can be transformed and thus result in other types (see Figure 6.1). His and subsequent work demonstrates two important phenomena in the relevance of design knowledge to design thinking. First, visual transformations within types are the result of formal transformations upon the class variables of the type. Secondly, new schemas may emerge through the transformation process of the original type. A new schema may be a substantive modification of the class variables and their associative relationships. This raises an interesting question regarding creativity in design: how can generic knowledge that is specific contribute to the emergence of new types?

3.2 Typology and topology

This model of emergence is based upon the interpretation of typological and topological knowledge. For example, in the case of a chair design the typology can be represented by various topologies and parametric associations among the basic components (Figure 6.1). Any modification can potentially result in the emergence of a new type.

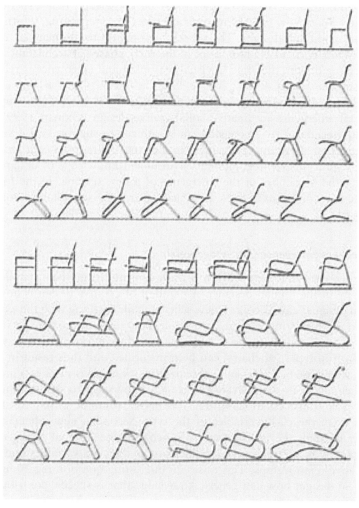

Figure 6.1 Topological and typological transformations according to Dickemann (1930).

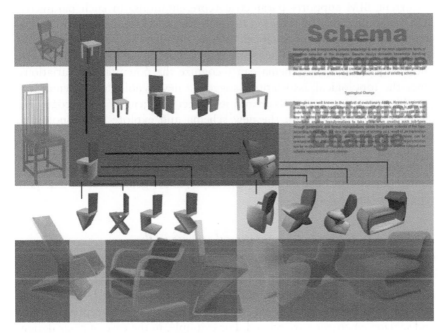

Figure 6.2 Schema emergence in design.

Source: Eyal Nir, Dani Brainin and Hezi Golan.

Figure 6.2 illustrates possible emergent types of a chair that are derived from the same objects. Emergence (morphogenesis) may be said to occur when the underlying structure is modified.

3.3 Schema emergence supported by BKM

Recently the technology of parametric design has been exploited to support topological manipulations and transformation in digital design. Most existing research on design emergence deals with shape emergence that has traditionally been modelled by instances. Instead of dealing directly with traditional models of transformations and shape manipulation we have attempted to understand how to model design transformations using parametric design.

Schema emergence can be supported by a BKM model. In this case shared knowledge would be related and negotiated via a shared representation. Parametric modelling has already been proposed as an effective means to embed domain expertise in models of BIM. These parametric-based approaches can exploit generic associative knowledge and thus provide domain-specific semantics and knowledge that can be shared by the various team members.

Current BIM supports parametric representations. In such parametric BIM versioning may be achieved through parametric transformations. The shared representational system can operate through the maintenance of the schema while enabling modifications and parametric variations. The generic knowledge behind the schema can recognize the potential for variations. Once the topological associations suggest a new typological schema, it is possible to explore a new type and its associated generic knowledge. The designer can interact with the new schema and explore new topological variations within a new typology.

This can be illustrated as follows: as the legs of the chair are expanded in the width dimension, they are transformed from a 'leg' object-type to a 'plane' object-type. The designer then explores the generics of an emerging type (see Figure 6.2).

3.4 Towards a shared representational schema in the computational environment

The emergence of a new schema is a fundamental cognitive capability of creativity in human collaboration. We can now demonstrate by an example how the concepts of visual design, design typologies, generic representations and schema emergence are related. For example, in the case of chair design, the typology of the chair can be represented by diverse combinations of sub-components. Within each particular element of the generic structure of the components of the type, modification is also possible as a means to differentiate the design image, and eventually results in the emergence of new sub-types. Transformations may be achieved through parametric variations, substitutions and other formal operations. Figure 6.3 illustrates analyses of possible schema emergence of a chair that is derived from the same graphical presentation of components. In the first case, transformations are achieved through parametric variations on components. In the second case, a sub-scheme of the first case, transformations are achieved through various formal operations upon the graphical components. Emergence may be said to occur in design when the underlying topological structure is modified.

This model of schema emergence has been a basis for the development of a computational environment that supports emergence. The computational environment provides an interactive design medium that is supportive of the cognitive capabilities of the designer. Schema emergence is supported by providing an interactive interface that assists in the construction of new design structures which can be derived from existing ones. The representational system operates through the maintenance of the schema while enabling modifications within the type. The generic knowledge acts in the background while the designer interacts with the representation dynamically to achieve transformations. Once the limits of transformation within a typological schema have been explored by the designer, it is possible to discover a new structure.

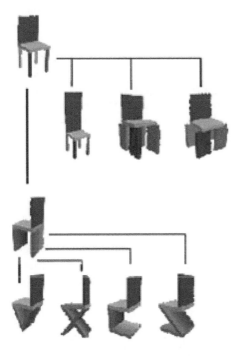

Figure 6.3 Formal analyses of schema emergence in a chair design which is derived from the same generic representation.

Source: Eyal Nir, Dani Brainin and Hezi Golan.

In order to support schema emergence in CAD, several problems and issues should be explored:

1 *Representation of knowledge behind the geometrical object.* How can typological and topological knowledge of the experienced human designer be represented in a CAD system? Current CAD systems describe object attributes which do not include typological knowledge of the geometrical object represented and associated variables. In order to do so, we need a theoretical basis to provide a representation of an object which would underlie the geometric properties of an object.

2 *Interpretation of generics within types and re-representation of new types.* How can a computational system interpret the generics within a current representation and re-represent a new representation of an emergent type once it has emerged?

3 *Interactive exploration of the type.* How can the associated modification of the new type be supported in a collaborative interactive environment?

In the following section we present a conceptual framework which supports the computational emergence of a new schema in design.

3.5 A computational framework for visual emergence

The following framework is defined by the three conditions above: knowledge representation, interpretation and exploration for emergence. The conceptual framework contains the following components, as illustrated in Figure 6.4:

1   *CAD interactive interface.* The environment should enable the designer to graphically manipulate the design object and to create instances within its own generic structure of representational possibilities.
2   *Linkages to the typological structure(s).* The typological definitions are those that support emergence in the CAD system, and the objects that implement the structure of the type class represent their generic definitions of elements, relationships and variables. In this case, concepts related to intelligent systems can be explored (Obonyo and Anumba 2007).
3   *Interpreter.* Our proposal is that the CAD objects should be linked to a mechanism which knows how to interpret the geometrical object as a schema: a simplified interpreter which activates a new type and its generics once one element of a different typological class has been instantiated in an exploration process. For example, in Figure 6.5, when the legs of the chair have been extended to the point in which they join to become planes, the new type emerges as a graphical representation which can be manipulated according to a new set of generics.
4   *Typological and topological interface.* An interface will allow us to see a structured instance in a visual mode. In fact, through the interface

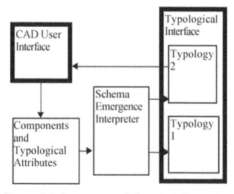

Figure 6.4 A conceptual framework supporting the emergence of a design schema.

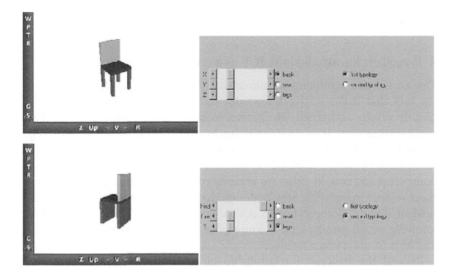

Figure 6.5 A typological interface – interpretation of a new typology related to parametric modifications.

we will allow the user to interact with the presentation and re-represent and define attributes and parameters, including their associated dynamic operations.

Figure 6.4 illustrates in an extended diagram how schema emergence may operate in collaborative design; here emergence is achieved by the human collaborators. Since we currently do not yet know how to implement a computational mechanism to support schema emergence, we propose a system architecture which allows the designer to make interpretations of a new schema, represent them and share them with his/her collaborator.

3.6 Implications

These ideas may have implications for creative collaboration in design. For example, the architect and engineer can communicate through a high-level representation of an emergent engineering schema as well as through an architectural schema and their interrelationships.

Since current systems do not support typological emergence, the following issue is raised: how to represent and interpret typological knowledge employing parametric techniques and associational relationships. In computational systems this is based on the capability of:

- the potential of a knowledge representation integrating both typology (object schema) and topology (geometry);

- interpretation of new typologies (object schema);
- recognition of emergent typologies (object schema).

4 Precedent knowledge: the ICF as a contextualized model

One of the characteristics of human-collaborative knowledge in design is the fact that collaborative knowledge is experiential and contextual and usually results as knowledge derived from the design process itself. This suggests that there is more in knowing how to design than just knowledge about factual information in a certain field of design. Human-collaborative knowledge in design is based on the fact that collaborative knowledge is contextual and usually relies on experiential knowledge. This suggests that there is more in knowing how to design than just knowing about the specific field of design. Competence in design praxis appears to be measured by knowing which specific kind of knowledge to apply in a particular situation and how to use it when needed.

Beyond the structured relationship of elements and associations that characterized the schema, there are other forms of conceptual knowledge that are derived from the experience of designing. In addition to the pattern of relationship formulated by the design scheme, there are other classes of knowledge that are of relevance to communication between team members. Significant among these formulations of design knowledge is the knowledge of *design precedents* (Oxman 1994, 2003). Design precedents are classes of design knowledge in which the formulation of knowledge has something to contribute to the future design problem. Thus learning to formulate precedent knowledge and to incorporate it in BIM would, for example, recognize the problem situation (for example, a conflict between architect and structural engineer) and be able to access the prior solution process. Furthermore, precedent-based formulations of knowledge might be a means whereby BIM systems could 'learn' from prior experiences.

The concept of the *design precedent* has had a significant role in design research. With reference to the domain of architecture, conceptual abstractions derived from the precedent are those which bridge between the conceptual and the physical and thus provide the *basis for exploiting the conceptual knowledge of precedents*. In our proposed approach for formulating precedent knowledge, the *acquisition and the construction of the body of concepts* from precedents are considered as means to demonstrate and facilitate the reuse of contextual knowledge. In a previous work (Oxman 2002) we argued that the formulation of the cognitive content of design thinking should be considered a main objective in design research. In order to implement one approach to this general objective, we have developed a framework in which knowledge acquisition is based upon the organization and development of contextual knowledge of design precedents.

The current version of this methodological framework supported collaborative design. The implemented computational system was called 'Web-Pad'.

In our view, in order to model design collaborative processes in a certain context, we can employ a conceptual mapping of related design ideas and concepts that can be constructed into larger structures. This framework provides the means to explicate knowledge of precedents and their contextual knowledge.

4.1 How to contextualize conceptual knowledge

The way to contextualize conceptual knowledge is as important as the amount of knowledge one possesses. This view emphasizes the notion of knowledge as a complex structure of contextualized concepts. Our conceptual structures, or the structure by which we organize our knowledge of the world, is not something of which we are naturally aware, and the formalization of such structures of knowledge requires an effort of semantic construction.

Think-Maps is a graphical and textual method that provides the means to formulate and structure the knowledge acquired by designers and can be employed in order to make it explicit. Think-Maps is methodologically based on a representational formalism called ICF originally developed for the representation of conceptual knowledge of designs (Oxman 1994, 2003). ICF is an organizational schema of knowledge. It was first developed as a computational model rooted in the theory and method of case-based reasoning. It employs a 'story' formalism that represents chunks, or independent segments, of conceptual knowledge that are intrinsic to design descriptions. In the Think-Maps framework, ICF acts as a structuring ontology for the construction of *conceptual networks of design concepts*.

4.2 Brief review of the ICF model

The ICF model is based upon a decomposition of holistic precedent knowledge into separate chunks. A design chunk, termed a 'story', is an original annotation of an entity of conceptual content that characterizes a specific design situation. A typical ICF chunk provides explicit linkages between issues of the design problem, a particular solution concept, and a related form description of the design, or a part of it.

A design issue is domain-specific semantic information related to the goals and issues of the problem class. Issues can be formulated by the programmatic statement, intrinsic problems of the domain, or the designer him/herself. A design concept is a domain-specific formulation of a solution principle, rather than the explicit physical description. A design form is the specific design artefact that causes the solution principle to materialize. For example, orientation is an architectural issue, centrality is an architectural concept to achieve orientation, and a specific engineering design may represent the actual formal and structural realization of this set of issues and concepts. A single issue may be addressed by different concepts, just as a single concept may address different issues especially among designers from different disciplines.

Among the objectives of the ICF model is to identify and represent individual components of design knowledge in a design in order that larger bodies of knowledge can be created from individual cases through a process of network construction. A further attribute of the semantic network structure is the ability to identify linkages between design ideas that were not originally apparent but can be established by navigating the connections between related design ideas. For example, Figure 6.6 illustrates an ICF conceptual structure of linked issues, concepts and forms. Each linked ICF presents a unique design idea.

The relationship between issues, concepts and forms in the ICF model is represented as a tri-partite schema. This schema has implications for memory organization, indexing and search that provide operative characteristics that are inherent to the network. Think-Maps is a repository collection of the design rationale based upon a structure of connected ICF modules as relevant to a design, engineering or construction situation.

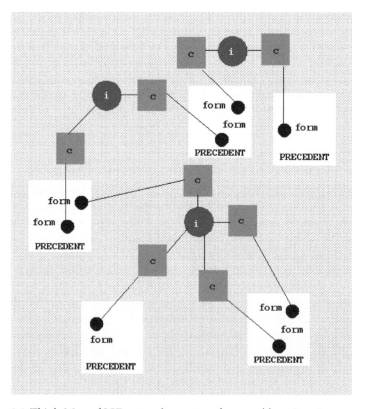

Figure 6.6 Think-Map of ICF network structure for a problem situation.

This concept can be applied to collaborative design. For example, see the text in Figure 6.7. The text represents the climatic design knowledge rationale embedded in the Green Tower in Dubai (see http://www.inhabitat.com/2007/01/30/o-14-dubai-commercial-tower/), designed by architects Jesse Reiser and Nanoko Umemoto (Figure 6.8). It demonstrates how knowledge is shared among the members of the collaborative team: the architect, the engineer and the environmental designer. This can be easily formulated as an ICF structure (Figure 6.9).

The building's façade perforations letting in light, air, and views through to the interior occupants. The one-meter space between the façade and the building's glass surface also yields a chimney effect causing hot air to rise, creating an efficient passive cooling system. The façade also serves as a structural exoskeleton, absorbing all of the tower's lateral forces and acting as a physical barrier for the building's window wall.

Figure 6.7 Textual description componentized into knowledge chunks.

Figure 6.8 The Green Dubai Tower (RUR Architecture, 2007).

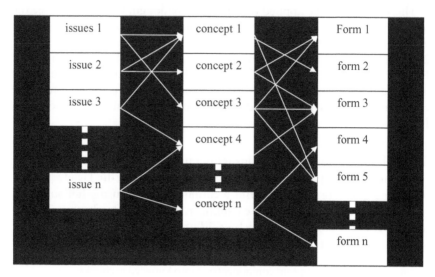

Figure 6.9 Explicating the conceptual mapping of domain knowledge.

4.3 The conceptual content of precedents and its application in BKM

Interdisciplinary collaboration among architects and engineers as form makers, and environmental engineers and manufacturers as agents for materialization are constructing new collaborative relations in the AEC professions. In such relations, informed processes such as performance-based design may generate a variety of intricate and complex relationships between the parts of the building, the agents of the total process, and the dynamic stages of the life cycle sub-processes (Oxman 2006). A BIM is a means to amplify the factual information and relationships in a dynamic model.

According to Augenbroe (2007), a sustained research effort has been devoted to achieve integration of input and output data from different applications in the architecture, engineering and construction (A/E/C) industry for the last 20 years. This has led to the introduction of incrementally richer versions of a BIM. However, developments in building product models have focused on a single uniform view. Furthermore, most product models and standards assume that all information about a building design is well structured and stored in 'structured idealizations' of reality and that data mapping can be automated. This ignores the need for an expert-driven knowledge, based on schematization skills and engineering judgement (Eastman 1999). Following this view, we believe that knowledge should be captured and employed to support experiential knowledge such as design judgement and the design rationale in collaborative work.

BKM can become a means to exploit precedent-based knowledge as a medium for formalizing such complex relationships between dynamic information structures and the diversity of the disciplinary rationale.

5 Conclusions

The function of a BKM has been proposed as one of the foundations to support collaboration among team members in digital design. Two cognitive models of knowledge that support collaboration were presented in this chapter.

The structure of data is constantly changing throughout creative collaboration in digital design. In human reasoning it is the function of knowledge that supports the modification of information structures which enable thought, innovation and creativity. A broad question emerges from this brief attempt to explore the potential of knowledge media as computational formalisms that might be incorporated within BIM. That question is: what kinds of knowledge are implicit in disciplinary scenarios that underlie the processes of the BIM, and how can this knowledge be formulated in order to be able to achieve BKM?

Meta-knowledge is the knowledge of how to organize what one knows. Currently BIM is often associated with data structures such as: Industry Foundation Classes (IFCs) and aecXML, which are data structures for representing information used in BIM. IFCs were developed by the International Alliance for Interoperability. There are other data structures which are proprietary, and many have been developed by CAD firms that are now incorporating BIM into their software. One of the earliest examples of a nationally approved BIM standard is the American Institute of Steel Construction (AISC)-approved CIS/2 standard, a non-proprietary standard with its roots in the UK.

Currently there is a need to develop meta-knowledge structures that go beyond such current BIM data structures and may help us to organize and exploit typologies in BIM technology, the way that human designers do.

In the first model, schema emergence appears to be a unique and highly significant form of emergence. The ability to preserve the schema of relationships even in conditions of modifications required by one agent of the team (for example, the structural engineer requires a modification in the architectural model) would be a medium for the communication of the design rationale to the members of the team. Furthermore, such a formalization of design knowledge can also potentially enhance the creativity of the team by supporting schema emergence. Thus the integration of the design schema as a form of meta-knowledge in BIM appears to be a valuable contribution. One possibility is to employ concepts of the Semantic Web (Obonyo and Anumba 2007).

Regarding the ICF model, we are at the beginning of the development of the idea of a BKM. As a generic definition, a BIM is a structured graphical and textual information relationship that is capable of accommodating diversity in content, scale, time and rationale of users. As such it is by definition a dynamic model and one capable of being updated during the various stages of design from conception to facility management in digital media.

How BKM should include these and other types of knowledge as a basis for collaborative design seems to be at the heart of a possible future for the idea of a BIM.

References

Augenbroe, G. (2007). 'Applying Process Rigor to the use of BIM in Building', in G. Q. Shen, P. S. Brandon and A. N. Baldwin (eds), *Collaborative Human Futures*. Hong Kong: Hong Kong Polytechnic University, Faculty of Construction and Land Use.

Eastman, C. M. (1999). *Building Product Models: Computer Environments Supporting Design and Construction*. Boca Raton, FL: CRC Press.

Huang, T., Kong, C. W., Guo, H. L., Baldwin, A. and Li, H. (2006). 'Virtual Prototyping System for Simulating Construction Processes', *Automation in Construction*, 16(5), pp. 576–585.

Obonyo, E. A. and Anumba, C. J. (2007). 'Towards a Multi-Agent Approach for Knowledge-Based Virtual Construction', in G. Q. Shen, P. S. Brandon and A. N. Baldwin (eds), *Collaborative Human Futures*. Hong Kong: Hong Kong Polytechnic University, Faculty of Construction and Land Use.

Oxman, Rivka (1994). 'Precedents in Design: A Computational Model for the Organization of Precedent Knowledge', *Design Studies*, 15(2), pp. 141–157.

Oxman, Rivka (1999). 'Visual Emergence in Creative Collaboration', in A. Brown, M. W. Knight and P. Berridge (eds), *Architectural Computing from Turing to 2000: The 17th ECAADE Conference Proceedings*, pp. 357–363. Liverpool, UK: Liverpool University.

Oxman, R. (2002). 'The Thinking Eye: Visual Re-cognition in Design Emergence', *Design Studies*, 23(2), pp. 135–164.

Oxman, Rivka (2003). 'Think-Maps: Teaching Design Thinking in Design Education', *Design Studies*, 25(1), pp. 63–91 (based on funded research).

Oxman, Rivka (2006). 'Theory and Design in the First Digital Age', *Design Studies*, 3, pp. 229–265.

Riese, M. (2007). 'Update on the Swire Properties Hong Kong One Island East Tower', in G. Q. Shen, P. S. Brandon and A. N. Baldwin (eds), *Collaborative Human Futures*. Hong Kong: Hong Kong Polytechnic University, Faculty of Construction and Land Use.

Schevers, H., Mitchell, J., Akhurst, P., Marchant, D., Bull, S., McDonald, K., et al. (2007) 'Towards Digital Facility Modelling for Sydney Opera House Using IFC and Semantic Web Technology', *ITcon*, 12, pp. 347–362, http://www.itcon.org/2007/24

7 Towards a multi-agent approach for knowledge-based virtual construction

E. A. Obonyo and C. J. Anumba

Abstract

In recent years, there has been a growing trend towards the use of virtual construction applications for design and functionality analysis in early stages of the project development cycle. It is the contention of this chapter that, although significant strides have been made in refining the capabilities of such applications, there is still no overall integration scheme for the sharing of information between the existing tools. This chapter proposes the use of an agent-enhanced knowledge framework within virtual construction applications. This will not only help address some of the information integration issues but will also make the modelling of virtual models more intuitive and powerful. The chapter gives an overview of the multi-agent paradigm, followed by a case study of a proof-of-concept agent-based construction e-business system (APRON) demonstrating the potential for using the multi-agent paradigm to address the challenges inherent in integrating information within distributed applications. This culminates in a discussion on the applicability of this approach in virtual construction applications.

1 Introduction

The overarching aim of this chapter is to demonstrate the potential for using the multi-agent paradigm to enhance the functionality of virtual construction applications. In recent years, there has been a growing trend towards the use of virtual construction applications for design and functionality analysis in early stages of the project development cycle. The motivation for using virtual design mock-ups is based on the need to increase constructability during the early phases of a project. Virtual mock-ups offer the project team a way of ensuring that any prefabricated components have been designed for assembly. Some very specific benefits of using such mock-ups include: 1) improving the visualization of the building model; 2) representation of multi-dimension design space through supporting the addition of new components or linking with various application packages; 3) providing real-time interactions among the design team, thus enabling the exploration of alternative ideas and design plans produced in a real design process; 4) providing multi-user real-time collaboration for problem solutions; and 5) linking with a broad range of information and accessing various databases or different domain application models through a network (Ding *et al.* 2003).

Although significant strides have been made in refining the functionality of virtual construction applications, there is still no overall integration scheme that addresses the challenges inherent in information management. This chapter discusses an information integration architecture based on the use of agent-based technologies. This will not only help address some of the information integration issues but will also enhance the functionality of virtual models. The proposed approach is based on an exemplary scenario: resolving design inconsistencies among a disturbed team developing virtual design mock-ups for prefabricated forming systems for concrete structures.

Multi-agent systems have been used by several researchers to build several complex construction-industry-specific systems ranging from e-business applications and cross-disciplinary communication to supply-chain management to contract administration. The chapter offers a case study of a construction e-business prototype that was implemented as a proof of concept and discusses how the approach can be adapted and deployed within a collaborative virtual construction context. Information agents, service agents, platform agents and application agents are the key components of the proposed knowledge-based framework for virtual construction. Some components of the proposed framework build upon existing knowledge in e-business, and XML and Web services standards. For the purposes of this chapter, the discussion is restricted to demonstrating how these technologies can be used collectively to execute requisite translation of data, information and knowledge into XML documents by intelligent agents, thereby creating the foundation for knowledge representation and exchange by other agents that support collaboration.

2 An overview of the multi-agent paradigm and the Semantic Web

It is not easy to define the term 'agents'. Nwana (1996) provides a number of explanations for this difficulty: it is a common term in everyday conversation; it encompasses a broad area; it is a meta-term; and researchers in this area have come up with such synonyms as 'bots', 'spiders' and 'crawlers'. Software agents in this project have been explored from the viewpoint of leading researchers. Brustoloni (1991), Maes (1994), Lieberman (1997), Jennings and Wooldridge (1998), Jennings *et al.* (1998), Ferber (1999) and the FIPA Architecture Board (2001) have defined the term 'agent' in various ways depending on their interests. It is, however, possible to extract some common attributes of agents from these definitions. There is a general consensus that software agents exist in an environment. They can sense the conditions in the environment, and such senses may affect how they act in the future. Software agents are also perceived as adaptive components that are capable of learning. They are proactive, exhibiting goal-directed behaviour. The execution of tasks occurs autonomously (without human intervention). Based on an analysis of these attributes, the term 'agents' as used in this chapter very loosely refers to systems capable of autonomous, purposeful action in the real world.

There are a number of closely related paradigms that need to be distinguished from agents. These paradigms include conventional programs whose output has no effect on what is sensed in the future. Such programs also lack temporal continuity (Franklin and Graesser 1996). Process control systems and software demons exhibit some features of agent-based systems, but they lack intelligence and flexibility. Agent technology bears close resemblance to other paradigms such as artificial intelligence, systemics, distributed systems, expert systems, remote programming and object-oriented programming. Such similarities are not surprising given that such techniques underpin agent technology (Aylett *et al.* 1997; Kashyap 1997; Fingar 1998; Jennings and Wooldridge 1998; Mahapatra and Mishra 2000).

The successful implementation of an optimal online system for the specification and procurement has been impeded by the existence of semi-structured or non-structured product information held in catalogues in various formats. Consequently, specifiers and procurers spend a significant amount of time gathering relevant information. This case study is based on a project aimed at demonstrating how agent technology can be used alongside other paradigms such as XML to make the specification and procurement of construction products more effective and more efficient.

The expansion in the use of the Web and its exponential growth are well-established facts. However, a problem has emerged from the very core of the success of the Web. The Web has evolved into a very large, unstructured but ubiquitous database holding over one terabyte of textual data (Baeza-Yates 1998). Such information overload on the Web has resulted in substantial losses in the modern economy. The results of one study established that in the US employees spend an average of eight hours each week retrieving external information (KPMG Consulting 2000).

The problems related to the rapid expansion in the use of the Web are also evident at the individual company level. There were great expectations that the use of Internet-enabled systems would improve productivity and reduce cycle times in various operations by collecting and providing the right information. A significant number of the forecasted gains remain unattained, partly owing to the lack of interoperability between systems (Madhusudan 2001; Samtani 2003). Collaboration has subsequently become more important for any company seeking to derive maximum gains from e-business (Deloitte Research 2002).

There is an evident need for Internet-based technologies that address the availability of information, the ability to exchange it seamlessly and the ability to process it across different applications in different organizational units. Consequently, the e-business world has been growing at an unprecedented pace, and new market realities, such as the software agent paradigm, have emerged. This chapter is based on an on-going project that explores the role of agents in construction-industry-specific e-business. The project focuses on developing an agent-based system for the specification and procurement of construction products on the Internet.

Significant research efforts have gone into achieving the Web described by Berners-Lee (1989). A number of researchers affiliated to the W3C are keen to advance the present functionality of the Web to a level where it will hold direct links to the information contained in the documents displayed on the Web. The W3C's Web of the future, which will hold machine-processable information, has been very broadly defined as the Semantic Web. It will allow people to find, share and combine information more easily (Hendler and Parsia 2002). W3C has set out to define and link the Web in such a way that it can be used for more effective information discovery, automation, integration, and reuse across applications. Hendler and Parsia (2002) identified the four key enabling technologies for the Semantic Web as:

- XML – adds arbitrary structure;
- RDF – provides a common framework for representing metadata across many applications;
- ontologies – store formal definitions of relations among terms; and
- software agents – automate tasks.

Clearly, software agents occupy a very important position as far as enhancing the efficient use of the Web is concerned. Consequently, a number of researchers have been able to establish the potential of using agents in e-business operations (Blake 2002; Blake and Gini 2002; Samtani 2003). The potential contribution of software agents to the success of e-business initiatives lies in their ability to collect Web content from diverse sources, process the information and exchange the results with other programs (Berners-Lee *et al.* 2001).

It is evident that software agents are growing in importance and will gradually become indispensable components of the Web of the future. There have been several significant research efforts across Europe. Such initiatives include the substantial investment by the European Union (EU) in large-scale, interdisciplinary and cross-country schemes such as Agentcities (http://www.agentcities.org/) and AgentLink (http://www.agentlink.org/). However, it is important to maintain realistic expectations. Leading researchers have admitted that there are still no large-scale implementations of agent-based applications in commercial scenarios (Luck *et al.* 2003; Shehory 2003). Agent technology is still maturing, and it will take at least another decade before there is a perfect match between the expectations and implemented systems.

3 Case study: agent-based information management in construction e-business

This section describes APRON (agent-based specification and procurement of construction products), a prototype agent-based framework supporting e-business through automating the specification and procurement of construction products. The work in the APRON project emerged from a partnership between Loughborough University and an industrial partner whose core business was hosting a construction-industry-specific e-business portal. The latter's main interest was creating a central repository of information about the built asset which allowed users to access information exactly tailored to their needs. At the start of the APRON research, the industrial partner had defined customized 'intelligent' components in AutoCAD that could be used to execute parametric searches for construction products. The implemented application scenario emerged from the need to provide a more flexible interface between these components in AutoCAD and the heterogeneous repositories provided by construction products' manufacturers.

3.1 The APRON architecture

The collaboration challenges in the case study are similar to those outlined in the general business context outlined in section 1. The processes in Internet-enabled specification and procurement of construction products involve retrieving information from autonomous units. Firms in the construction industry are under increasing pressure to improve their ability and capacity for collaborating with members of their supply chain. This makes it imperative for collaborating firms to have interfaces to the applications used by the team members across corporate boundaries.

APRON can be perceived as an e-business support system hosted by a construction-industry-specific information portal. It comprises components that execute search functions, manage content, enable collaboration and manage the processes involved in the specification and procurement of construction products. The first basic assumption in the research was that targeted manufacturers have made the required information on construction products available on the Internet. It was further assumed that the participating parties operate autonomously as far as their information systems are concerned. In the implemented prototype, the agent-based framework enabled the final end-users to have direct access to information published by manufacturers without significant changes being made to their existing computing applications. The very essence of APRON is facilitating the reuse of legacy components in their existing form.

3.1.1 The functional components

The APRON prototype is modelled using the established mediation/wrapper methodology that was used in the InfoSleuth prototype (Bayardo *et al.* 1997) and the SEEK prototype (O'Brien *et al.* 2002). The final prototype comprises a software middle layer between the semi-structured repositories of construction products and the end-user applications utilized in specification and procurement. This is depicted in Figure 7.1. There are three distinct, intercommunicating layers. The construction product manufacturers are at the top level in the architecture. Manufacturers display details of the various product offerings as semi-structured data on the Internet. The kernel of the architecture is the e-marketplace of an information provider, who uses the Internet to ensure that requisite project information is available to all the key players in the construction project supply chain. This e-marketplace hosts the APRON solution, which consists of the download, extraction, structuring, database, search and procurement modules. The e-marketplace also holds a repository of relevant standards that can be used as XML schemas for structuring the extracted information. The final layer comprises the end-user firms in the specification and procurement of construction products. The focus in these firms is providing an automated interface to the computing applications used by specifiers and procurers.

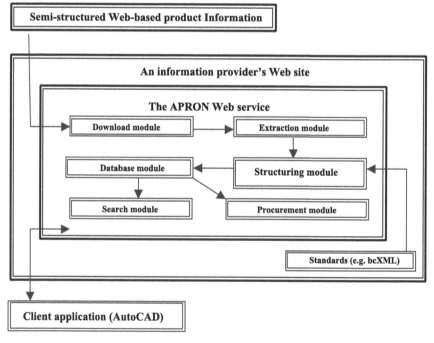

Figure 7.1 The implemented APRON prototype system.

A client application has been deployed to allow such end-users to communicate with the APRON Web service.

The APRON system provides a link between the Web site holding product information and the applications used in the specification and procurement of construction products. The download module maintains real-time access with these Web sites. Text is then extracted from the downloaded file and, using a template of previously defined XML schemas, structured into a context-specific format. Industry standards, such as ifcXML and bcXML, can be easily adopted and used to create the template used for structuring the extracted information. The APRON system also stores the relevant information in a database. Information for the specification and procurement of construction products is obtained from this database. APRON provides a Web-based search engine, which can be used to execute context-specific queries for construction products. The APRON solution also offers a framework for automating the procurement of specified products. The procurement module has two types of agents: a buyer agent and a seller agent, representing product procurers and product sellers respectively. The two agents exchange requisite information and automate the transactions involved in the procurement of construction products.

3.1.2 Deployment scenario

The example implemented in the APRON prototype focused on processing product information for the specification and procurement of light bulbs from the Philips Lighting Web site (http://www.lighting.philips.com/). The site hosts close to 200 catalogues in Adobe Acrobat PDF format. The information that would be of interest to an end-user (for example, wattage, cap size and voltage) is presented in a semi-structured format. The Web site does not have a search facility that would support guided navigation based on attributes such as wattage and voltage. It is also not possible to query the information directly from any another application. Furthermore, relevant data has to be re-keyed for reuse elsewhere. It is therefore necessary to educe the contents of the light bulb table from the underlying Web page, displayed in Table 7.1.

APRON offers support for the e-marketplace to create an extended enterprise at two levels: 1) supporting a designated portal administrator; and 2) supporting an 'ordinary' end-user. Web portal solutions already package their solutions in a manner that restricts access to certain 'sensitive' capabilities to designated portal administrators. Within the APRON framework, access to the download, extraction, structuring and database modules is restricted to an authorized portal administrator. These are Java-based modules implemented as software agents using X-Fetch Suite's AgentServer as a mediator/facilitator.

The APRON system ensures that the data is as current as possible. A designated administrator presets the download, extraction and structuring

Table 7.1 Light bulb information source

Commercial product name			
Type	Wattage	Cap/base	Voltage
MASTERline ES	20W	GU5.3	12V
MASTERline ES	25W	GU5.3	12V
MASTERline ES	30W	GU5.3	12V
MASTERline ES	35W	GU5.3	12V
MASTERline ES	40W	GU5.3	12V
MASTERline ES	45W	GU5.3	12V
MASTERline ES	50W	GU5.3	12V

modules to execute at preferred intervals. The administrators must execute the initial download, extraction and structuring cycle before the end-users present their first request for information to avoid delays in response time. The URL for the target Web site has been hard-coded within the download module. This is based on an assumption that many specifiers and procurers of construction products have preferred suppliers with whom they have established a working business relationship over time. Should there be a desire to access information from other Web sites, an administrator can easily broaden the relevant agent's roaming habits on the Internet. The scope of the download module is specifically delineated to handling product data displayed in the Adobe Acrobatic PDF format. It was also established through an informal review of various Web sites that a significant number of construction product manufacturers use the PDF format to display information on the Web. The scope of APRON can be easily broadened to accommodate many other formats, such as HTML.

The extraction module is an optional module included in the implemented prototype because the target PDF files on the Philips Lighting Web site were not in ASCII format. The extraction module educes all the text from an identified target source. This is fed into the structuring module, which utilizes the capabilities of the X-Fetch Suite to structure the extracted information into XML format. The extraction module parses the source file based on a predefined schema based on the data extraction language (DEL). The extracted text is registered and stored as element names or tags to be used in a structured XML document. Figure 7.2 shows an example of the code for the extraction module. The structuring module then generates an XML file using the extracted column titles as element tags and the product details as values. An excerpt of the resulting XML file is depicted in Figure 7.3.

Evoking the agent responsible for the structuring module a second (and any subsequent) time updates the original XML output file if the product manufacturer has made modifications on the source file. It is important to note that the XML file would have been created regardless of the source data

```
<mark action="go"/>
<extract type="over" expression="Commercial product name"/>
<extract type="re_upto" expression="[a-zA-Z]"/>
<extract type="upto" expression=" " register="title_1"/>
<extract type="re_upto" expression="[a-zA-Z]"/>
<extract type="upto" expression=" " register="title_2"/>
<extract type="re_upto" expression="[a-zA-Z]"/>
<extract type="upto" expression=" " register="title_3"/>
<extract type="re_upto" expression="[a-zA-Z]"/>
<extract type="upto" expression=" " register="title_4"/>
<extract type="re_upto" expression="[a-zA-Z]"/>
<extract type="upto" expression=" " register="title_5"/>

    <convert conversionset="toXML">
    <register name="title_1"/>
    <register name="title_2"/>
    <register name="title_3"/>
    <register name="title_4"/>
    <register name="title_5"/>
      </convert>
        <repeat>
```

Figure 7.2 Extracting product attribute values.

```
<product>
        <Type>CAPSULEline Pro</Type>
        <Wattage>10W</Wattage>
        <Cap_base>G4</Cap_base>
        <Voltage>12V</Voltage>
        <Ordering>409706 50</Ordering>
    </product>
    <product>
        <Type>CAPSULEline Pro</Type>
        <Wattage>20W</Wattage>
        <Cap_base>G4</Cap_base>
        <Voltage>12V</Voltage>
        <Ordering>402103 50</Ordering>
    </product>
    <product>
        <Type>CAPSULEline Pro</Type>
        <Wattage>20W</Wattage>
        <Cap_base>GY6.35</Cap_base>
        <Voltage>12V</Voltage>
        <Ordering>402196 50</Ordering>
    </product>
```

Figure 7.3 The output XML file.

format. The DEL reasons on different file formats in the same way. For this code to work using a different source file, in a different format and from a different manufacturer, it is imperative to have consensus on the key attributes that define a selected construction product. Such consensus can be arrived at through the existing standardization efforts. However, the existing standards are still developing and are not yet robust enough (Froese 2003).

Once product data has been presented in a structured format, the APRON system provides a component that allows administrators to create permanent data stores of the information. The information that is downloaded from manufacturers' Web sites is stored only temporarily on the server. The designated administrators have authorization to create and update databases for the structured data. MS Access was used for APRON's database module. This component can be easily adjusted to accommodate other types of databases.

The APRON proof of concept was implemented to explore how the functionality of the industrial partner's AutoCAD-based 'intelligent' components could be further enhanced. AutoCAD was therefore used as the end-user application in the pilot scenario. All the components necessary to support end-user interactions run on the APRON Web service shown in Figure 7.1. The client application for AutoCAD was packaged into a visual basic application (VBA) that interfaces with the APRON Web service using a dynamic link library. Loading and running the VBA from AutoCAD presents the end-user with a form, shown in Figure 7.4, that can be used to search for detailed product specifications using known attribute definitions.

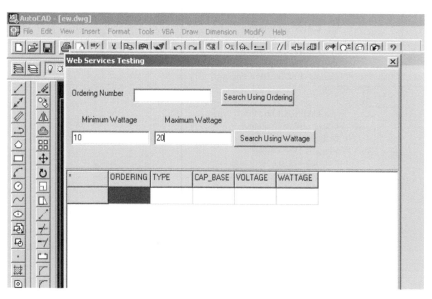

Figure 7.4 AutoCAD-based search form.

4 Multi-agent support for information management in virtual construction

A key benefit to using virtual mock-ups during the design phase is that this approach can bring together parts or aspects of design being explored by different members of the project team. This gives distributed designers an opportunity to ensure that the different components of the design are compatible through detecting and resolving such things as design mismatches. Clearly, the development of virtual construction mock-ups involves the various participants collating information from different sub-systems. It is therefore expected that, for one to deploy large-scale applications, they would have to address information integration. The problem of managing vast volumes of information is actually something that always has to be addressed in systems focused on securing IT-enabled team collaboration. The chapter proposes the use of the agent paradigm to address this challenge. The proposed framework is based on the agent-based approach to digital management information adopted in the APRON case study as outlined in the preceding section.

The proposed approach also builds on an existing agent-based framework for virtual prototyping within the context of manufacturing to support the life-cycle collaboration roundly and effectively (Yaoqin *et al.* 2006). An adapted view of this framework has been presented in Figure 7.5.

Ideally, virtual construction applications should encapsulate the views of participants from different domains using specialized software applications including the various CAD packages, estimating and scheduling software, project planning software and enterprise resource planning (ERP) systems,

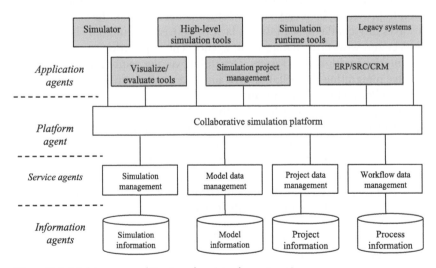

Figure 7.5 Multi-agent architecture for virtual construction.

Source: Adapted from Yaoqin *et al.* (2006).

as shown in Figure 7.5. Because of the reliance on historical data from previous jobs, it may be necessary to incorporate a mechanism for incorporating data from several legacy applications. In addition to these challenges, projects typically involve several participants from conceptual design to detail design, manufacturing of prefabricated components, executing job-site operations, commissioning and using the facility and decommissioning it at the end. This creates a need for dynamism in any system based on information representing the full life cycle of any constructed facility.

Section 2 defined multi-agent systems. These systems build on distributed artificial intelligence. Their robust capabilities are based on intelligent agents with specialized task assignments cooperating to attain a common goal. Agent-based applications are therefore ideal for scenarios such as virtual construction prototyping which inherently require a team of distributed experts to collate their information to solve a problem. Clearly, there is a potential for using the agent paradigm to enhance the seamless flow of information in the generation of virtual mock-ups.

Figure 7.5 shows that the proposed approach is based on the use of four types of agents: information agents, service agents, platform agents and application agents. The information agents: 1) represent such things as simulation information, model information, project information, workflow information and resource information; 2) handle integrity and consistency of information; and 3) control access to information (Yaoqin *et al.* 2006). Virtual simulations are used in construction to check design mismatches through such things as calculating the critical path or project duration and harmonizing the workflow to minimize idle time during the construction phase. The service agents represent these management-support services. The platform agent is the control node of the multi-agent community. In addition to providing directory service for the other agents in the system, it also resolves any conflicts, such as the lack of consistency in goals between the project and workflow management agents. The applications agents provide the interface with the end-users. Depending on the requirements, this may involve integrating the collaborative simulation platform with their custom tools and/or computing applications.

It was previously established that the motivation for using the multi-agent approach is based on the agents' ability to cooperate in distributed problem-solving. The example in Figure 7.6, based on the need for simulating workflow in a project, shows how this goal can be achieved. The relevant application agents propose a service request to platform agents, which in turn evoke the service agents that interoperate with information agents to access, achieve, gather or transfer distributive information. The information is then communicated back to the platform agents.

Clearly, this level of cooperation will not be possible without an ontology for the domain. The proposed approach can be based on the use of industry standards such as IFCs. IFCs essentially offer a framework for representing information for architecture, engineering, construction and facility

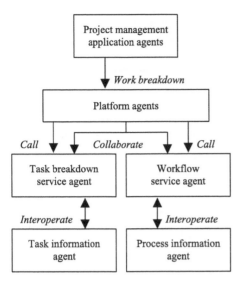

Figure 7.6 Cooperation among agents.

Source: Adapted from Yaoqin *et al.* (2006).

management projects. The IFC object descriptions deal with not only full 3D geometry but also relationships, process, material, cost and other behaviour data. Figure 7.7 depicts an exemplary taxonomy that can be used to define ontology for the collaborative generation of virtual mock-ups for prefabricated forming systems.

5 Conclusion and further work

Although significant strides have been made in refining the capabilities of virtual construction applications, there is still no overall integration scheme for the sharing of information between the existing tools. This chapter has presented a conceptual framework for using the multi-agent paradigm to support the seamless flow of information in such applications. The proposed approach is based on the use of an approach that had been used for information integration within construction e-business and demonstrated through the implementation of a proof-of-concept system (APRON). The chapter gave an overview of the multi-agent paradigm, followed by a case study of the APRON prototype. It also discussed how the underlying principles can be adapted for use in virtual construction applications. The chapter has discussed how the multi-agent paradigm can be used to make virtual construction models more intuitive and powerful.

Further work in the research will involve developing a proof-of-concept system to demonstrate the potential for using the multi-agent paradigm to address the challenge of information integration within the context of a

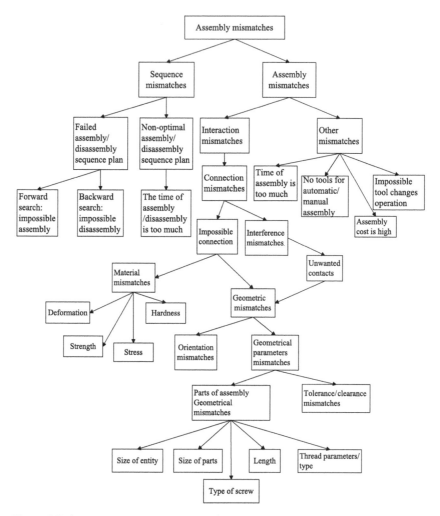

Figure 7.7 A taxonomy to support virtual construction.

Source: Adapted from Taratoukhine and Bechkoum (2000).

virtual construction application. The approach proposed in this chapter will be used in an exemplary scenario to demonstrate how intelligent agents can be used to enhance the benefits that can be derived from use of virtual mock-ups for prefabricated temporary structures such as forming systems.

References

Aylett, R., Brazier, F., Jennings, N., Luck, M., Nwana, H. and Preist, C. (1997). 'Agent Systems and Applications', Report from the panel discussion at the Second UK Workshop on Foundations of Multi-Agent Systems (FoMAS '97).

Baeza-Yates, R. A. (1998). 'Searching the World Wide Web: Challenges and Partial Solutions', in H. Coelho (ed.), *Progress in Artificial Intelligence: IBERAMIA 98, 6th Ibero-American Conference on AI*, Lecture Notes in Computer Science, 1484, pp. 39–51. New York: Springer.

Bayardo, Jr., R. J., Bohrer, W., Brice, R., Cichocki, A., Fowler, J., Helal, A. et al. (1997). 'InfoSleuth: Agent-Based Semantic Integration of Information in Open and Dynamic Environments', *Proceedings of the 1997 ACM SIGMOD International Conference on Management of Data*, 11–15 May 1997, Tucson, AZ, pp. 195–206. New York: ACM.

Berners-Lee, T. (1989). *Information Management: A Proposal*. Geneva: CERN.

Berners-Lee, T., Hendler, J. and Lassila, O. (2001). 'The Semantic Web', *Scientific American*, May, http://www.sciam.com (accessed 15 January 2002).

Blake, M. B. (2002). 'AAAI-2002 Workshops: Agent-Based Technologies for B2B Electronic Commerce', *AI Magazine*, 23(4), pp. 113–114. AAAI Press.

Blake, M. B. and Gini, M. (2002). 'Guest Editorial: Agent-Based Approaches to B2B Electronic Commerce', *International Journal on Electronic Commerce*, 7(1), pp. 5–6.

Brustoloni, J. C. (1991) *Autonomous Agents: Characterization and Requirements*, Carnegie Mellon Technical Report CMU-CS-91-204. Pittsburgh, PA: Carnegie Mellon University.

Deloitte Research (2002). 'Directions in Collaborative Commerce: Managing the Extended Enterprise', Deloitte Research Collaborative Commerce Viewpoint, Deloitte Research Report.

Ding, L., Liew, P. S., Maher, M. L., Gero, J. S. and Drogemuller, R. (2003). 'Integrating CAD and 3D Virtual Worlds Using Agents and EDM', in M.-L. Chiu, J.-Y. Tsou, T. Kvan, M. Morozumi and T.-S. Jeng (eds), *Digital Design: Research and Practice*, pp. 301–323. Amsterdam: Kluwer.

Ferber, J. (1999). *Multi-Agent System: An Introduction to Distributed Artificial Intelligence*, pp. 9–11, 53–56. Harlow, UK: Addison-Wesley.

Fingar, P. (1998). 'A CEO's Guide to E-Commerce Using Object-Oriented Intelligent Agent Technology', CommerceNet Research Report, No. 98–19, USA.

FIPA Architecture Board (2001). 'FIPA Agent Software Integration Specification, 2001', http://www.fipa.org (accessed 30 September 2002).

Franklin, S. and Graesser, A. (1996). 'Is It an Agent, or just a Program? A Taxonomy for Autonomous Agents', in M. J. Wooldridge and N. Jennings (eds), *Intelligent Agents III: Agent Theories, Architectures and Languages (ATAL)*, pp. 21–35. Berlin: Springer.

Froese, T. (2003). 'Future Directions for IFC-Based Interoperability', *ITcon*, 8, Special Issue: 'IFC – Product Models for the AEC Arena', pp. 231–246.

Hendler, J. and Parsia, B. (2002). 'XML and the Semantic Web', *XML Journal*, 3(10).

Jennings, N. R. and Wooldridge, M. (1998). 'Applications of Intelligent Agents', in N. R. Jennings and M. Wooldridge (eds), *Agent Technology: Foundations, Applications, and Markets*, pp. 3–28. Berlin: Springer.

Jennings, N. R., Sycara, K. and Wooldridge, M. (1998). 'A Roadmap of Agent Research and Development', *Autonomous Agents and Multi-Agent Systems*, 1(1), pp. 7–38.

Kashyap, N. (1997). 'An Expert Systems Application in Electromagnetic Compatibility', Master's thesis, University of Missouri-Rolla, USA.

KPMG Consulting (2000). 'Knowledge Management Research Report', http://www.kmadvantage.com/docs/km_articles/KPMG_KM_Research_Report_2000.pdf (accessed 3 July 2003).

Lieberman, H. (1997). 'Autonomous Interface Agents', in S. Pemberton (ed.), *Proceedings of the ACM Conference on Computers and Human Interface*, pp. 67–74. Reading, MA: ACM/Addison-Wesley.

Luck, M., McBurney, P. and Preist, C. (2003). 'Agent Technology: Enabling Next Generation Computing – A Roadmap for Agent-Based Computing, Version 1.0', AgentLink Report.

Madhusudan, T. (2001). 'Enterprise Application Integration: An Agent-Based Approach', IJCAI Workshop on AI and Manufacturing, Seattle, WA, August.

Maes, P. (1994). 'Agents that Reduce Work and Information Overload', in J. M. Bradshaw (ed.), *Software Agents*. Menlo Park, CA: AAAI Press/MIT Press.

Mahapatra, T. and Mishra, S. (2000). *Oracle Parallel Processing*, 1st edn, Chapter 1. Farnham, UK: O'Reilly.

Nwana, H. S. (1996). 'Software Agents: An Overview', *Knowledge Engineering Review*, 11(3), pp. 205–244.

O'Brien, W., Issa, R., Hammer, J., Schmalz, M., Geunes, J. and Bai, S. (2002). 'SEEK: Accomplishing Enterprise Information Integration across Heterogeneous Sources', *ITCON – Electronic Journal of Information Technology in Construction*, Special Edition on Knowledge Management, 7, pp. 101–124.

Samtani, G. (2003). *B2B Integration: A Practical Guide to Collaborative E-Commerce*. Hackensack, NJ: World Scientific Publishing.

Shehory, O. (2003). 'Agent-Based Systems: Do They Provide a Competitive Advantage?', *AgentLink Newsletter*.

Taratoukhine, V. and Bechkoum, K. (2000). 'An Agent-Based Paradigm to Support Virtual Design Mock-Ups', ECAI 2000: 14th European Conference on Artificial Intelligence Workshop: Applied Semiotics – Control Problems.

Yaoqin, Z., Huizhong, W., Xueqin, C. and Shenglei, C. (2006). 'Application of Multi-Agent System in Virtual Prototype Engineering for Complex Product', in K. Koyamada, S. Tamura and O. Ono (eds), *Systems Modeling and Simulation Theory and Applications*. Tokyo: Springer.

8 Update on the Swire Properties Hong Kong One Island East Tower BIM success story

Martin Riese

Introduction

The Swire Properties Limited One Island East (OIE) Tower design and construction project in Hong Kong is one of the most substantial building information modelling (BIM) implementations ever undertaken. The owner, Swire Properties Limited, commissioned Gehry Technologies (GT) to put into place a robust implementation of new BIM technologies and working methods to help to lower the cost of construction, by enhancing efficiency and reducing waste across the entire process.

The owner, architect, structural and mechanical engineers, quantity surveyors, Gehry Technologies, and project contractors and subcontractors collaborated to create a single 3D BIM project database. Using BIM, over 2,000 clashes were identified and resolved prior to tender. The resulting tender returns were lower and within 1 per cent of each other, owing to the enhanced quality of design and construction information.

The winning contractor, Gammon Construction Limited, assumed full responsibility for the BIM model during construction and ensured that all 2D information that went to the site was first vetted in the virtual 3D prototype. The project design and construction teams used the new technologies and working methods to greatly improve the process of coordination, cost control, 2D drawing production and construction methodology simulation.

During construction, the virtual prototype became the central management tool for identifying and coordinating thousands of clashes and construction sequence issues before previously unidentified problems reached the site. The internet-based 3D project database provided stakeholders, the project team and the entire project supply chain with instant real-time visibility in the entire building lifecycle process.

The OIE project was completed in May 2008. In addition to the shortened construction time, when the final project accounts have been completed Swire Properties Limited in Hong Kong is also anticipating the confirmation of a substantial saving on the construction cost resulting from the implementation of BIM. The project was awarded the American Institute of Architects 2008 BIM award for design/delivery process innovation.

The implementation of Digital Project software and GT working methods on this project have demonstrated the added value that is now achievable by design, construction and facilities management teams throughout the world. As this technology and its associated working methodology are adopted globally by the construction industry, universal improvements in safety, quality, cost and build time will be realized.

This chapter will review the implementation of various key aspects of the building lifecycle information management technology and working methods that delivered the OIE success. This project forms one of the most substantial 'work in progress' examples of the implementation of collaborative technologies and working methods in the construction industry as a whole.

In the beginning

In 2004, Swire Properties Limited in Hong Kong commissioned GT to implement and manage the BIM process for the OIE office tower project, which is located in Hong Kong.

Objectives

Swire Properties Limited and GT have been collaborating for nearly four years to bring 'construction industry transformation' to the Asia region. The initial objective was to save 10 per cent on the cost and six months on the time to construct this 70-storey office tower. Swire specified Digital Project throughout the detailed design and construction phases and into facilities management.

In general, project teams working with current collaborative technologies and working methods can anticipate the following benefits:

1 Comprehensive three-dimensional geometric coordination of all building elements, which in itself is widely accepted to achieve a reduction in construction cost of 10 per cent.
2 A fully integrated, web-based building information infrastructure that can be combined with web-based building management tools in addition to those already provided by GT.
3 An integrated collaboration infrastructure that tracks the allocation of tasks and performance of project participants in relation to 3D geometry and metadata (replacing traditional project portals). This infrastructure helps to manage and track the performance of tasks and task participants, and it also provides a knowledge base and search capabilities to correlate and compile information regarding building com-

ponents or issues encountered in construction. The use of e-mail is thereby eliminated on construction projects.

4 Automated identification, reporting and management of clashes and conflicts, allowing the designers to resolve conflicts before construction, resulting in a significant reduction of rework and waste.

5 Enhanced quantity take-off from the BIM to improve the speed and accuracy of the preparation of the bill of quantities before and after tender. Real-time, accurate, complete project cost knowledge – as often as is necessary.

6 Lower, more accurate tender pricing resulting from the contractors' unknowns and risks being identified, managed and reduced prior to tender.

7 Automation and two-way interoperability of two-dimensional documents with three-dimensional building information model data.

8 Direct integration into the unified building lifecycle management database environment of structural engineering analysis and simulation, energy analysis, fire safety analysis, and code compliance.

9 Fabrication-level capabilities such as automated shop drawings and direct supply chain integration.

10 Creation of a reusable catalogue of intelligent parametric building parts and project knowledge capture.

11 Management of construction sequence and process simulation using the Digital Project BIM elements. The BIM process is directly linked to the construction process modelling simulation. This provides an additional reduction in the cost of construction if implemented adequately by the constructor.

12 Reduction of waste in the construction throughout the entire process.

13 Order-of-magnitude reduction of contractor requests for information (RFIs).

14 Significant reduction of claims on-site resulting from incomplete design coordination.

15 Quicker construction.

16 Lower construction cost (10 to 30 per cent is achievable).

17 Standardization of information exchange throughout the entire construction supply chain and regulatory authorities enabled by IFC compliance.

18 Other benefits associated with Industry Foundation Classes (IFC)-compliant data exchange, such as automated 3D regulatory authority submissions.

19 Continuation of the industry-wide commitment to maximizing safety.

20 Better build quality.

21 Integrated, internet-based post-completion facilities management and maintenance using the BIM elements in concert with purpose-made facilities management integration tools.

Implementation

To start the project together, Swire and GT did the following:

- opened a common BIM project office;
- specified computers and software;
- organized an internal project team;
- provided offices and computers for the project team members;
- provided software and training for the architects, mechanical, electrical and plumbing (MEP) engineers, structural engineers, quantity surveyors and all contractors.

The OIE success story

On OIE, GT provided Swire Properties Limited with the following building lifecycle management (BLM) services:

- Digital Project training for the entire project team, including:
  - one week of Digital Project software basic training,
  - one week of Digital Project software advance training for MEP engineers and architects, and
  - one week of 'learning by doing' (full-time GT question and answer during modelling);
- developed BIM methodology for the project;
- trained and embedded a GT model manager on the project team;
- provided full-time architectural, structural and MEP BIM technical support for the project team;
- provided full-time quantity surveying BIM technical support for the project team (including scripting services to partially automate the formatting of the bill of quantities into Hong Kong Institute of Surveyors (HKIS) format);
- produced the initial BIM model and then transferred it to project team;
- trained all the bidding contractors to a level sufficient to query the BIM model and significantly reduce their risk – prior to tender;
- provided the winning contractor with full BIM training;
- provided the contractor with a trained GT model manager;
- provided ongoing technical support to the winning contractor;
- provided support to the contractor's 4D construction process modelling team;
- provided marketing materials for the owner directly from the BIM model;
- provided technical BIM support to the contractor's supply chain (such as Gartner);
- provided BIM process quality assurance to the owner;
- provided support for the transition of the contractor's BIM model back

to the Swire facilities management team and trained the owner's facilities management team;
- assisted Swire with deriving metrics for the actual value realized by this BIM implementation (ongoing);
- helped the owner to transfer the knowledge of this successful BIM implementation to other projects.

During the 12-month design development phase, a comprehensive BIM model was made by the OIE project design team.

Integration of project team collaboration

One of the key advantages of using a robust BIM platform such as Digital Project is that it enables concurrent building lifecycle information modelling (BLM) by large construction project teams, with virtually no additional software customization time or training (less than a day). Digital Project, with its integrated Concurrent Versioning System (CVS,) delivers a significant cost improvement over the implementation of traditional project data management (PDM) systems in this industry, which require months or even years of customization and implementation.

Specification of deliverables

The new technologies and working methods deliver significant return on investment at any stage of a project and with virtually any level of existing

Figure 8.1 The One Island East design team.

Table 8.1 Project information

Contract type	Competitive tendering
Construction cost	US$300 million (approximately)
Project scale	70 floors with two basement levels
	Total floor area: 141,000 m^2 (1,517,711 sf2)
	Typical floor area: 2,270 m^2 (24,434 sf2)
Current stage	Under construction
Owner	Swire Properties Limited
Architect	Wong & Ouyang (HK) Limited
Quantity surveyor	Levett & Bailey Quantity Surveyor Limited
Structural engineer	Arup Hong Kong
Contractor	Gammon Construction Limited
BIM consultant	Gehry Technologies
Function	Office
Structure	Reinforced concrete and some structural steel
Exterior	Aluminium curtain wall

information. In principle, BIM cannot be 'farmed out' to offshore 'modelling shops' as a stand-alone service. Effective BIM-enabled project teams collaborate directly in regular existing project meetings to identify clashes and unresolved design issues, and continually incorporate the modifications into the live BIM model. The BIM model is generally present in project team meetings during the design, construction and facilities management phases. Although building lifecycle information modelling is a never-ending iterative and collaborative process, it can be generally divided into the following overlapping phases:

- *Phase 1: architecture and structure BIM.* Creation of a building information model – based on the 2D and 3D project information available, including architecture and structure, which forms the framework on which to model and coordinate the MEP. Included are all elements of structure and architecture that have significant interfaces with MEP.
- *Phase 2: MEP BIM.* Creation of an MEP building information model. Included are all major MEP elements required to identify and resolve clashes in the construction information such as ducts, pipes, sprinklers, cable trays, lights and plant items. The model includes the identification of clashes between these items and openings in structure and architecture.
- *Phase 3: ongoing coordination of the design.* Ongoing iterative incorporation into the BIM of redesign resulting from clash identification and coordination of the design. This is an ongoing process that does not

stop until project completion. The purpose of this phase is to use the model created in Phase 1 and Phase 2 to bring the construction information to a state in which it can deliver lower tender prices and an improved construction and facilities management process.

- *Phase 4: construction.* After contract award, the contractor continues to benefit from the BIM process by pre-coordinating the detailed design for the project prior to construction. The contractor, assuming responsibility for the construction of the project, should use the BIM to check every aspect of the project prior to construction.

- *Phase 5: facilities management.* After completion, the BIM model is passed back to the owner as a highly accurate 3D 'as-built' record of the built works. The traditional boxes of binders of operating and maintenance (O&M) manuals are replaced by the BIM, which contains 'hyperlinks' to all of the information about the project. The building management system (BMS) and fire control system can be linked directly to the BIM model. Taking advantage of further 'hard wire' connections to all of the elements of the building, the BIM at this point makes the transition to being a full building lifecycle model (BLM) by providing two-way connectivity to the project – via the internet – to the owner's facilities management team.

BIM costs

Building information modelling is an ongoing process of iterative collaboration and construction information database management which continues to project completion and beyond. As such, it is difficult to provide a fixed price for a building information model. It is widely accepted that the implementation of the technology and working methods is less than 1 per cent of the construction cost, and almost incidental when compared to the lifecycle cost of running the building.

Construction BIM model

Essential to the realization of the maximum potential value of BIM is the transfer of the model to the contractor during construction. This is best achieved by making it a contractual obligation that the contractors adopt and continue to develop the BIM model and technology implementation as part of the creation of their construction information.

RFID

Another element that is key to the realization of the maximum potential of technology-enabled building lifecycle management is the radio frequency identification system (RFID). A minimum requirement of any BIM solution is that it provides integral RFID interoperability, enabling design-driven automated RFID number sequence changes and linkage to advanced supply chain management systems. Although RFID was not implemented on the OIE BIM project, the contractor encoded internet 'hyperlink' information into each element to connect it to its supplier. This technology will help project teams to enjoy full transparency in the entire supply chain for the project.

Industry Foundation Classes (IFC)

It is important that BIM solutions be IFC compliant and construction projects should be conducted using information that conforms to internationally accepted IFC standards. This will further enhance the building lifecycle information management value achievable though advanced interoperability. Although IFC is itself still developing, it can already enable improved information exchange and automated, internet-based, 3D regulatory authority submission processes.

Facilities management integration

The OIE BIM process has been passed from the BIM consultant, to the designers, to the constructors and then back to the owner's facilities management process. The elements of the BIM and 4D models produced by the project team will be used to enable the owner's implementation of advanced internet-based facilities management tools.

Procurement

The method of procurement selected by the owner for this project was the traditional tender process. GT trained a number of potential main contractors in the use of quantity extraction and measurement from the BIM model. During the process of tendering, the potential constructors detected a number of errors in the model, which were corrected prior to tender. The BIM model did not form part of the legal contract documents, but was appended to the contract 'for reference'. In future, the BIM model, with its superior accuracy, will itself become the contract document.

Tender returns

Providing the potential contractors with the 3D BIM model helped them to evaluate the level of risk and resulted in lower tender returns. The difference between tender prices was less than in a traditional 2D tender.

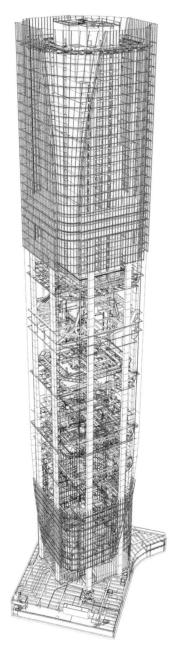

Figure 8.2 Cutaway drawing of the pre-tender design team BIM model.

Figure 8.3 Site progress in August 2007.

Main contractor selected

The successful main contractor was Gammon Construction Limited. The MEP subcontractor was Balfour Beatty. The cladding was provided by Gartner.

BIM transferred

Construction began in March 2006. GT provided the main contractor with three weeks of basic and advanced training. The main contractor assumed full responsibility for the tender BIM model and began to develop it into a highly accurate and detailed 3D construction BIM model. Gammon and Balfour Beatty were able to keep the progress of the OIE BIM model ahead of the construction at all times.

During construction

The construction BIM model became the main visualization tool for coordinating the many elements of the project. The main contractor maintained a team of eight full-time modellers who helped to identify clashes and coordination issues and incorporated the resulting design solutions back into the model. A number of subcontractors also modelled their elements of the works.

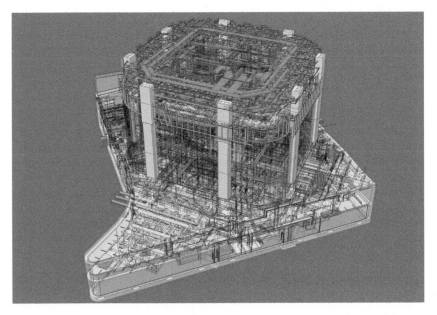

Figure 8.4 Gammon–Balfour Beatty basement to 3/F construction BIM model.

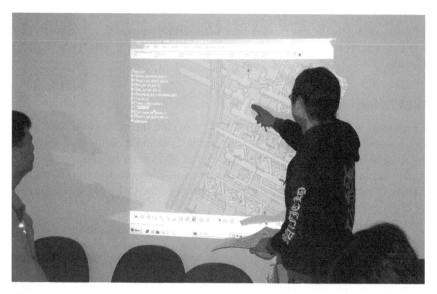

Figure 8.5 Image (1) taken in the contractor's design coordination sessions showing the integration of the BIM model into the construction process.

Source: Gammon Construction Limited.

Shop drawings

All shop drawings – including all MEP shop drawings – were reviewed against the design intent as indicated in the BIM model and then incorporated into it. Any requirements for revisions that were identified in this shop drawing process were returned to the relevant subcontractor for incorporation into revised shop drawings.

Combined builder's work drawings (CBWDs)

The 2D combined builder's work drawings used to place the elements of the work on-site were all 'vetted' in the 3D BIM model process prior to the placement of formwork and MEP elements on site. The result was a substantial reduction in abortive works. In future, the majority of CBWDs will be produced automatically from the BIM. Any 2D drawing which is used to communicate geometry to the construction site should be generated automatically from the BIM model, because it will contain far fewer errors than drawings made by people.

OIE 4D construction simulation

One of the major contributing factors to the success of many GT projects around the world – including the OIE project – has been the use of 4D

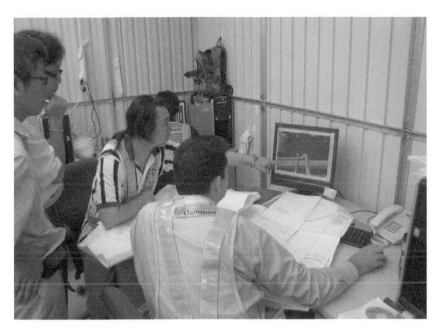

Figure 8.6 Image (2) taken in the contractor's design coordination sessions showing the integration of the BIM model into the construction process.

Source: Gammon Construction Limited.

simulation. Visualization is a simple graphic representation of a known outcome. Simulation, on the other hand, explores the effect on the outcome of various optimization scenarios. It is, therefore, a tool to help to find optimized methodologies that may not be possible without the use of the simulation. This is one example of how the implementation of new technologies can help to lead the industry to an order of magnitude of improved safety, quality and efficiency. Gammon Construction employed this tool extensively to help to significantly improve the construction sequence and to manage risk. Gammon Construction has said that construction process modelling saved the project 20 days.

Incorporating Gammon Construction's expertise with the power of process simulation, the contractor's team conducted detailed construction sequence optimization exercises before the actual construction. A number of sequencing problems and clashes were identified – particularly in critical sequences such as the four-day floor construction cycle and the erection of the outrigger floors. The construction process was optimized in this way, helping the team to realize the 24-month schedule. Job one – safety – is improved by reviewing people's tasks in advance.

Whilst geometric coordination of the design prior to construction is thought to yield a 10 per cent cost saving, construction process modelling

Figure 8.7 Detailed construction information for all elements of the work – including MEP and cladding elements – was coordinated and incorporated into
the BIM model prior to construction.

is thought to potentially contribute an additional 20 per cent cost saving on
the construction. The highly effective construction process modelling
employed by Gammon Construction Limited on the OIE project was produced by the Virtual Prototyping Laboratory at the Hong Kong Polytechnic
University, in cooperation with GT.

The 4D visualization was produced using Dassault Systemes' Delmia
software, incorporating the BIM elements produced by the project team
using Digital Project. To be effective, BIM solutions must exchange information seamlessly with time management and visualization platforms such as
Microsoft Project, Primavera and Delmia.

In the automobile industry, process simulation is thought to deliver up to
a 50 per cent value improvement. It is possible to imagine that, in the not-
too-distant future, the transformed industry of the built environment may be
able to deliver this kind of value to its clients and stakeholders. Process
optimization is likely, in itself, to open up design possibilities that are currently not achievable. Streamlining and integration of supply chain process
knowledge into preliminary design will not only improve efficiency but
accelerate the scope and rate of innovation.

Figure 8.8 Table form moved off formwork lift in OIE construction process model.
Source: Gammon Construction Limited and the Virtual Prototyping Laboratory at the Hong Kong Polytechnic University.

Figure 8.9 Actual table forms being moved on-site.

Outrigger construction

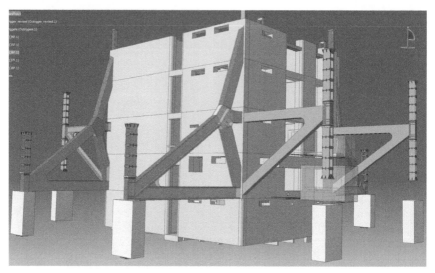

Figure 8.10 Outrigger in BIM model.

Figure 8.11 Outrigger on-site.

Cladding installation

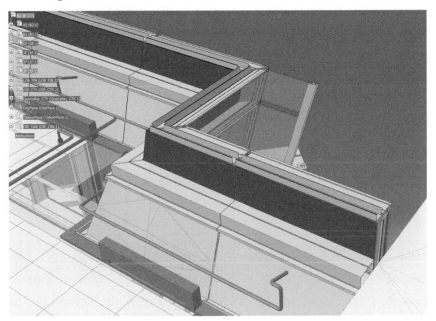

Figure 8.12 Cladding in the BIM model.

Figure 8.13 Cladding on-site.

Managing cost with BIM

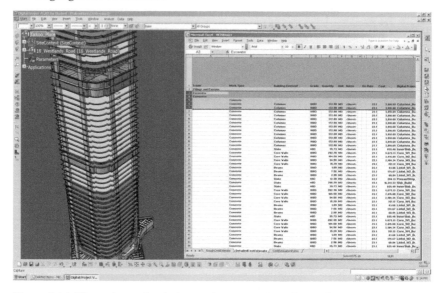

Figure 8.14 Throughout the design and construction phase, the BIM was used to monitor cost in real time. The technology can produce vast, detailed and appropriately formatted quantity take-off information in real time. This helps the entire project team to manage cost more effectively throughout the project.

Managing time with BIM

Figure 8.15 shows how the sequence of construction of each element of the project is part of the 'data tree' of the BIM model. This means that, through Primavera interoperability, sequences can be changed and the corresponding element numbers can automatically be revised in the BIM model. This functionality was used on OIE to visualize and simulate progress.

The sequence numbers are intended to correspond to RFID numbers which help to manage the flow of elements of the project from the construction industry supply chain to the site and into the facilities management phase.

Two-dimensional drawings

Existing BIM technology can now generate all general arrangement drawings automatically from the BIM model (Figure 8.16). The effort and resource formerly used in the construction industry to revise and embellish 2D drawings can now be directed to the task of enhanced 3D coordination. The days of manually cutting sections and coordinating 2D construction drawings are over. More importantly, this industry needs to go beyond paper-based processes as soon as possible.

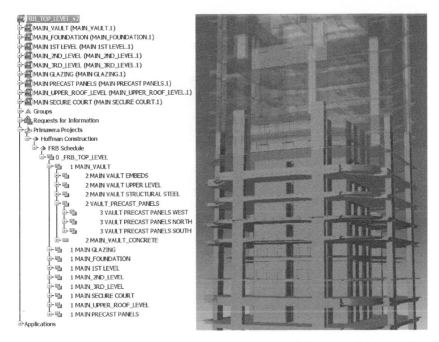

Figure 8.15 The sequence of construction becomes part of the BIM information that defines the project.

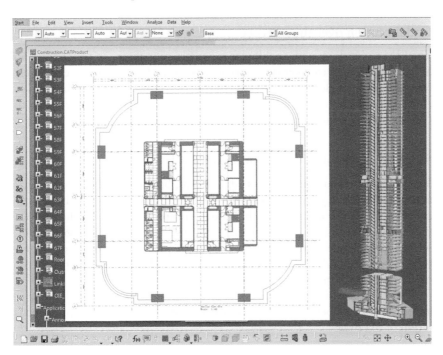

Figure 8.16 Building information modelling enables automated 2D drawing production from the 3D database.

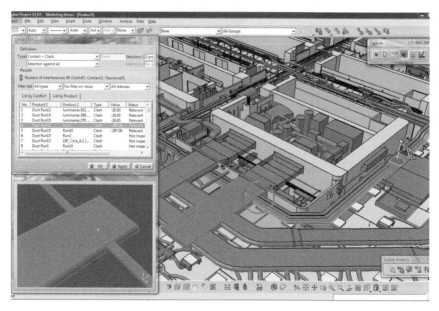

Figure 8.17 Automated clash detection and automated file management were essential elements of the OIE BIM process.

Collaboration over the internet

The BIM technology now enables large construction teams to collaborate on a single BIM database over the internet. Team members can work effectively together on a single project across time zones, continents and disciplines. Automated clash detection and file management bring increased freedom to concentrate on design, coordination and optimized procurement.

Swire Properties and OIE project metrics

Swire Properties management has advised that, because the OIE project final accounts have not yet been completed, the metrics have not yet been fully collated. However, evidence of the value delivered is ample on the project. Bids were lower than expected owing to the increased clarity of construction information. There were far fewer RFIs on the project – an order of magnitude less than would be expected. The project is believed to be below the original budget. The project is on time.

This value is typical of other GT projects throughout the world and is in keeping with – or better than – established global metrics for the implementation of 3D BIM.

Conclusion

The OIE project was completed ahead of time in March 2008. In addition to this time saving, Swire Properties Limited in Hong Kong is anticipating a saving on the construction resulting from the implementation of BIM. The comprehensive implementation of Digital Project and GT working methods on this project has demonstrated the added value that is now achievable by design and construction teams everywhere. As this technology and working methodology is adopted globally by the industry of the built environment, substantial improvements in safety, quality, cost and build time will be realized by everyone.

Acknowledgements

Portions of this chapter, including images, appeared previously in the following publications:

'Trends in Building Lifecycle Information Management', published in the November 2007 issue of *Indian Architect and Builder*. Special thanks to *Indian Architect and Builder* for permitting their use in this document.

Southeast Asia Building magazine, March/April 2006 issue. Thanks are due to SEAB for permission to reprint parts of the article.

The Proceedings of the Australasian Universities Building Educators Association (AUBEA) Conference 2006. Special thanks are due to the conference organizer for permission to reprint parts of the article in this paper.

Tse, T. C., Wong, K. D., Wong, K. W., Leung, Y. M. and Riese, M. (2006). 'Building Information Modelling: A Case Study of Building Design and Management in One Island East, Hong Kong', *Proceedings of the Australasian Universities Building Educators Association (AUBEA) Conference 2006*, 11–14 July 2006. Sydney: University of Technology Sydney.

Virtual Futures for Design, Construction and Procurement, edited by Peter Brandon and Tuba Kocatürk (Wiley-Blackwell, 2008).

9 IKEA pattern and revolution of the construction industry

*Heng Li, H. L. Guo, T. Huang,
Y. K. Chan and M. Skitmore*

1 Introduction

Construction management and technology are two key factors influencing development of the construction industry. In the past 40 years, although some new and advanced technologies have been applied to construction projects, the efficiency of the industry is still in decline. The main reason for this is that the new technologies do not effectively reduce the cost of design and construction, especially for design/build (D/B) projects, or improve the management of construction. For example, although computer-aided design (CAD) technology has improved the efficiency of drawing, the number of design errors, and therefore the amount of rework, remains unchanged. CAD also offers little prospect of reducing costs through the optimisation of construction processes or enhancement of management decision making. What is needed, therefore, is both 1) appropriate new technology and 2) a change in current management concepts to improve the efficiency of the construction industry.

The IKEA organisation is well known for its efficient management and manufacturing approach for home furnishings. With suitable technological adaptation, the approach offers a new way for the management and construction of construction projects. In the following sections, the IKEA approach is briefly introduced and the adaptations necessary for use on construction projects are described. Finally, a case study is described to illustrate its use in practice.

2 The IKEA approach and construction projects

IKEA is a famous brand of home furnishings, and its vision is that 'IKEA offers a wide range of well-designed, functional home furnishing products at prices so low that as many people as possible will be able to afford them' (IKEA website). The purpose of the IKEA approach is to fulfil this vision. This involves designers working with manufacturers to find smart ways to make furniture while still using existing production processes. Consumers choose the furniture themselves, pick it up at the self-serve warehouse, and then assemble it easily using the 3D instructions provided (an example is shown in Figures 9.1 and 9.2) without any professional help. This means

Figure 9.1 An IKEA dining table.

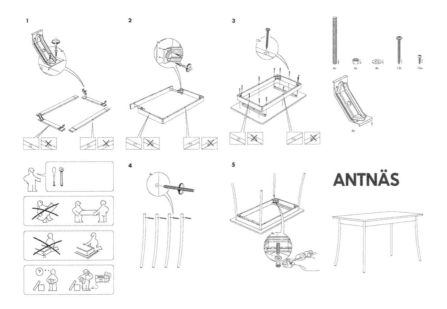

Figure 9.2 The 3D assembly instructions for an IKEA dining table.

that IKEA incurs no assembly costs and is therefore able to reduce the sale price accordingly. Note that the appropriate manufacturing sequence is the basis of the 3D instructions.

In theory at least, the IKEA approach is also suitable for construction projects, especially design and build/construct projects. If a construction project can be simplified sufficiently, just like a table, it may also be easily constructed based on 3D instructions.

In order to make appropriate 3D instructions for a construction project, two key issues need to be addressed: 1) the design has to be error free, and 2) an appropriate construction sequence has to be devised. However, the designs and construction sequences of construction projects are quite different from those of home furnishings. Construction projects have no fixed design plans and construction sequences. With current technology, it is far too costly to check the design or manufacturing sequence of a project in the same way as for home furniture. Instead, what is needed is a virtual environment in which design plans and construction sequences can be tried again and again to develop a design and appropriate construction sequence. In addition, the design of a construction project involves many specialist activities, for example architecture, structure, building service (BS), temporary support and so on. Thus collaborative design among different parties is also needed. To do this on a virtual platform involves the use of virtual prototyping (VP) as the key technology.

3 Collaborative design

3.1 Virtual prototyping

Virtual prototyping (VP) is a computer-aided design process concerned with the construction of digital product models (virtual prototypes) and realistic graphical simulations that address the broad issues of physical layout, operational concept, functional specifications, and dynamics analysis under various operating environments (Pratt 1995; Xiang *et al.* 2004; Shen *et al.* 2005). Dedicated VP technology has been extensively and successfully applied to the automobile and aerospace fields (Choi and Chan 2004). For instance, an automobile can be fabricated virtually via the VP technology, which allows various team members to view the 3D image of the finished products, evaluate the design, and identify production problems prior to the actual start of mass production.

VP technology has also been applied to some extent in the construction industry in the form of construction process simulation, with the Construction Virtual Prototyping Laboratory (CVPL) of Hong Kong Polytechnic University being a prominent leader in the field (Huang *et al.* 2007). It is believed that VP can be used to check the design efficiently, enable rapid modifications, and then simulate the construction process in a virtual environment so as to present a clear and easily operated 3D construction instruction. For construction projects, this involves the use of two software products, Catia V5 and Delmia V5, both belonging to Dassault Systemes.

3.2 Design error checking

The design of construction projects is much more complicated than that of home furnishings because of the many participants involved and the great deal of information required. As a result, design errors often occur during the process of design or the period of construction. In order to successfully and conveniently construct a building using IKEA-like 3D instructions, these errors must be eliminated before the project is begun. Through the 3D main model of the project, design errors can be found easily. The procedure of checking design errors is presented as follows: 1) construct a 3D main model using Catia V5, including architecture, structure and BS, for example the 3D model of a spectator stand (see Figure 9.3); and 2) automatically detect clashes between all components in the model (see Figure 9.4), for instance clashes found in the model of the spectator stand (see Figure 9.5).

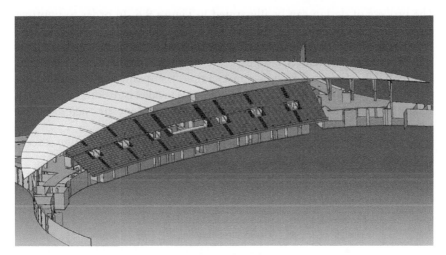

Figure 9.3 The 3D model of a spectator stand.

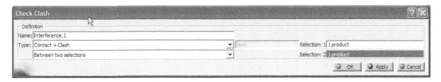

Figure 9.4 Checking design errors automatically.

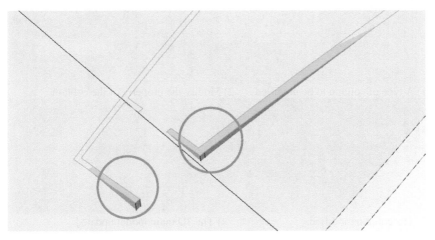

Figure 9.5 Clashes between roof and partitions.

After finding the design errors, all participants can discuss and modify them in the 3D main model.

3.3 Modification of design

The process of design can be illustrated in Figure 9.6. In order to improve the efficiency of design, all the design errors detected must be modified rapidly. In the virtual environment, it is easy to modify these errors using Catia V5. When any element is modified, the 3D main model can be updated automatically. Figure 9.7 shows an example for modifying a column. To modify the length of this column type all that is needed is to edit the property 'length' of the column. All instances of this column type are then modified automatically in the 3D main model.

An appropriate and detailed design plan is then developed through several cycles of design modification.

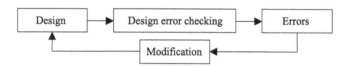

Figure 9.6 The process of design modification.

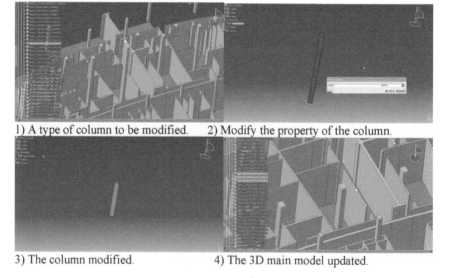

1) A type of column to be modified.   2) Modify the property of the column.

3) The column modified.   4) The 3D main model updated.

Figure 9.7 The rapid modification of design errors.

4 Construction process simulation

The core idea of the IKEA approach is to provide 3D instructions for the site workmen. The 3D instructions must consist of a detailed and appropriate construction sequence. As mentioned earlier, the construction sequence of a construction project cannot be repeated in a real environment and must therefore be tested in a virtual environment. This is the aim of the construction process simulation, and its procedure is shown in Figure 9.8. Based on the design, the simulation of the construction process is conducted in a virtual environment using the Delmia V5 software product.

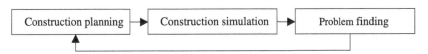

Figure 9.8 The process of construction planning simulation.

4.1 Test of construction sequence

Through simulating different construction sequences using Delmia, their feasibility can be analysed, including collisions, conflicts, safety and so on. As an example, Figure 9.9 shows the simulation of the construction sequence of a V-column installation. Here, it is obvious that the temporary support is outside of the first-floor slab and therefore the construction sequence is not feasible. Similar instances can also be identified and the construction sequences modified until all of these problems are solved.

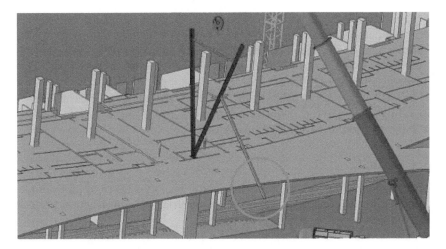

Figure 9.9 The simulation of a V-column installation.

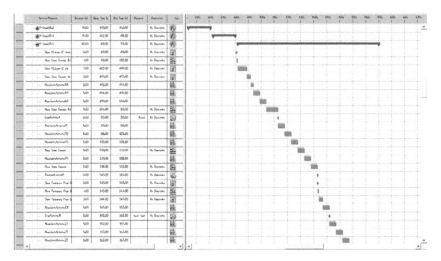

Figure 9.10 Schedule of a V-column installation.

4.2 Optimisation of construction sequence

Feasible construction sequences, including resources levelling, sequence reordering and so on may be optimised further. By adjusting sequencing, construction time may be reduced and, via resources levelling, construction cost saved. Figure 9.10 shows the schedule for installing a V-column.

By testing and optimising the construction process, an appropriate construction plan is eventually developed. This provides the guideline for compiling the IKEA-like 3D construction instructions.

5 The IKEA approach to construction project management

Based on the above-mentioned design errors checking and construction process simulation, an appropriate design and construction plan is identified. Following this, the IKEA approach to construction project management becomes possible, namely 3D construction instructions are developed. The process is illustrated in Figure 9.11.

The 3D construction instructions are step by step and very clear. Therefore, by referring to the 3D instructions, workers are able to work smoothly without the need for any other technical help.

6 Case study

The V-column installation of a construction project is taken as an example to illustrate the application of the IKEA approach to construction project management.

Firstly, related 3D models should be established, including V-columns, columns, slab, partition and so on (see Figure 9.12) using Catia. Secondly,

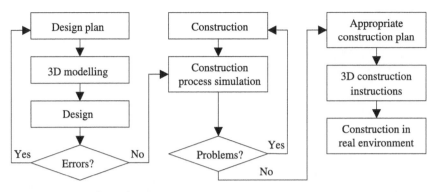

Figure 9.11 The flow of applying the IKEA pattern to construction projects.

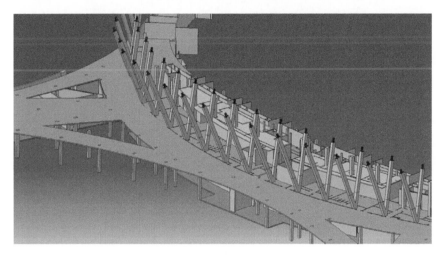

Figure 9.12 3D models of components related to the V-column.

design errors are checked, for example the clash shown in Figure 9.13. Thirdly, these design errors are modified. The design errors are eventually eliminated after repeated design-checking modifications. Fourthly, the process of installing the V-column is simulated with Delmia based upon the construction plan needed to make the process feasible. Some problems are found, for example the conflict shown in Figure 9.9. Fifthly, the construction process is modified after discussions with subcontractors. When all problems are solved, an appropriate process of installing the V-column emerges. Finally, the 3D instruction for the V-column installation is compiled, as shown in Figure 9.14.

As with the installation instructions for IKEA furniture, it is clear and easy for anyone to install the V-columns using the associated 3D instruction. This makes the installation process one of low cost and high safety. The development of the 3D instruction for the V-column installation indicates the

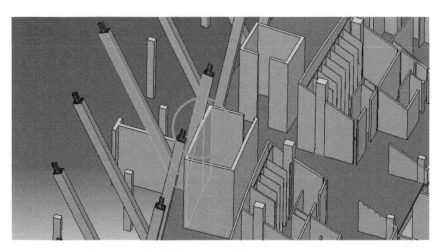

Figure 9.13 A clash between a V-column and partition.

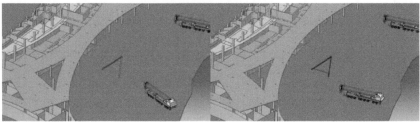

1) Weld the UC columns of V-column. (2) Install temporary steel channel.

Figure 9.14 The 3D instruction of the V-column installation.

IKEA approach with VP technology (Catia and Delmia) to be feasible for construction projects.

7 Conclusions

The new technology and management methods have led to gains in the efficiency of the manufacturing industry but have not been possible in construction owing to the often bespoke nature of construction projects. In order to enhance the efficiency of construction projects, the IKEA approach, as an efficient manufacturing and management means in the home furnishings field, may be introduced through adopting VP technology. Two key factors, design without errors and appropriate construction sequencing, were identified as being necessary for the implementation of the IKEA approach in construction project management using two VP software products, Catia and Delmia. The implications on the construction process were then summarised. From the case study it was shown that, when used in conjunction with VP, the IKEA approach may be suitable for use in construction projects to better instruct workers in their work.

It is believed that the application of the IKEA approach in construction project management will improve the efficiency of construction projects and enhance the degree of safety of its workers. It is likely that design and build/construction projects, in particular, will be most amenable to the approach, which has the potential to revolutionise the way in which construction projects are undertaken.

Acknowledgement

Professor Peter Brandon has been very helpful and provided important guidance to the work of the Construction Virtual Prototyping Laboratory.

References

Choi, S. H. and Chan, A. M. M. (2004). 'A Virtual Prototyping System for Rapid Product Development', *Computer-Aided Design*, 36, pp. 401–412.

Huang, T., Kong, C. W., Guo, H. L., Baldwin, A. and Li, H. (2007). 'A Virtual Prototyping System for Simulating Construction Processes', *Automation in Construction*, 16(5), pp. 576–585.

Pratt, M. J. (1995). *Virtual Prototyping and Product Models in Mechanical Engineering: Virtual Prototyping – Virtual Environments and the Product Design Process*. London: Chapman & Hall.

Shen, Q., Gausemeier, J., Bauch, J. and Radkowski, R. (2005). 'A Cooperative Virtual Prototyping System for Mechatronic Solution Elements Based Assembly', *Advanced Engineering Informatics*, 19, pp. 169–177.

Xiang, W., Fok, S. C. and Thimm, G. (2004). 'Agent-Based Composable Simulation for Virtual Prototyping of Fluid Power System', *Computers in Industry*, 54, pp. 237–251.

10 Improving information delivery

Jeffrey Wix

The business case

Virtually all of the surveys carried out in the building construction industry place interoperability as a key issue for the use of information and communication technologies (ICT). This is supported by road map developments such as that carried out for the European Construction Technology Platform (ECTP) and supporting national efforts. The evidence for available cost benefit comes from a study by the US National Institute for Standards Technology (NIST) (Gallaher *et al.* 2004) in which the lack of interoperability was estimated in 2004 to cost US industry more than US$15 billion per year. This equates roughly to 1.5 per cent of US construction turnover. However, more recent estimates suggest that this figure is too low, the true figure being closer to US$45 billion or 4.5 per cent of turnover (European Construction Technology Platform 2005).

Interoperability and information delivery

Wikipedia defines 'interoperability' as 'connecting people, data and diverse systems', allowing the term to be defined either in a technical way or in a broad way, taking into account social, political and organizational factors (http://en.wikipedia.org/wiki/Interoperability). It is a widely used but frequently abused term. In this chapter, the term 'information delivery' is used instead since it is considered to be more directly relevant to what the industry needs to achieve and what it needs to improve.

In looking at improving information delivery, the focus is not only on data but also on the software applications (systems) involved in delivering (and receiving) information and the actors (people and organizations) concerned with delivery. Most importantly, it is concerned with the business processes across which information is delivered and the extent to which the delivered information can be used in managing the execution of business processes.

The key to information delivery

The key to information delivery is that there must be a common understanding of the building processes and of the information that is needed for and results from their execution. For the information, this is provided by the Industry Foundation Classes (IFC) information model that provides a comprehensive reference to the totality of information within the lifecycle of a constructed facility (International Alliance for Interoperability 2006). It has been developed since 1996 as an integrated whole in response to business needs identified by the international building construction community and is now mature, powerful and stable.

The complete IFC information model is developed as a set of individual topic schemas. Each topic schema typically represents a consistent overall idea (e.g. structural analysis, HVAC, cost, materials, etc.). On completion, all of the topic schemas are brought together into the single schema which is the authorized working version. The core part of the IFC model (termed the 'platform') forms the ISO/PAS 16739 standard (ISO 2005). This is shown as the light grey items in Figure 10.1.

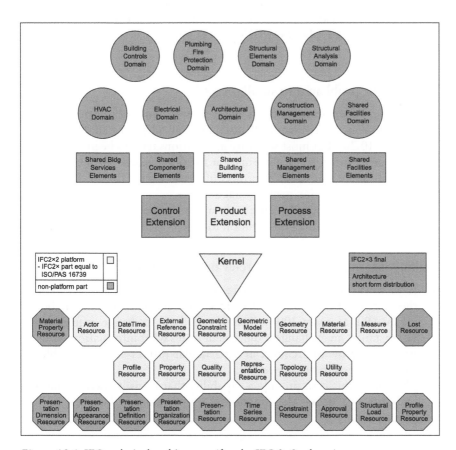

Figure 10.1 IFC technical architecture (for the IFC 2×3 release).

IFC can store information about most aspects of a building project, including:

- shape;
- schedule;
- cost;
- client brief and requirements;
- design and analysis information;
- procurement and construction data;
- operation and maintenance records.

Whilst use of the IFC model is happening successfully on 'real world' projects in many places, it has been found that uptake is improved when the business needs of practitioners are fully supported in the IFC-based information exchange.

In response to this, the concept of the Information Delivery Manual (IDM) has evolved within buildingSMART. It is being further defined as ISO 29481–1, which is expected to be complete and available in 2009.

The following describes how IDM works and how it supports the use of the IFC model.

Breaking the problem into small pieces

- The IFC information model is huge. It deals with all information in a construction project from all disciplines at all points in the lifecycle (Figure 10.2a). But most software applications tend to be more specific. They deal with a set of information for one purpose at one point (or possibly a few points) in the project lifecycle (Figure 10.2b). This is also what the construction professional does.
- That means they need to use only a few of the capabilities that IFC offers. The IDM offers the solution to this.

Making information exchange relevant to the user . . .

- IDM works by cutting out a piece of the IFC model that is relevant to the user, and hence to the software vendor supporting the user (buildingSMART Norway 2007). This is called an 'exchange requirement'. For instance, an exchange requirement states what information needs to be provided by an architect at a particular project stage to enable an energy analysis to be carried out.
- The aim of this is to provide a capability that is easily used in industry and that can define the 'handover' of information between project participants.

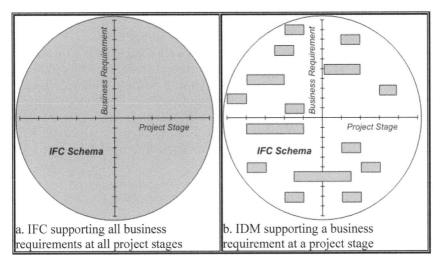

Figure 10.2 IDM support for business processes.

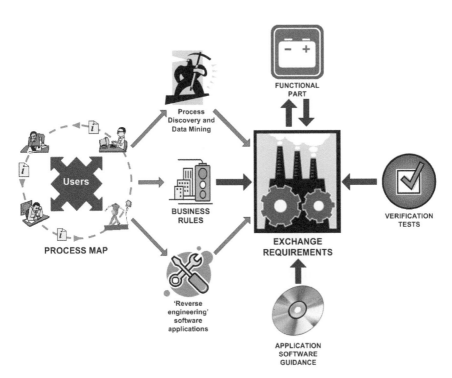

Figure 10.3 Exchange requirement factory.

... and to the software provider

- An exchange requirement is the user-facing part of IDM.
- However, there is also a software-provider-facing part that is termed a 'functional part'. A functional part provides the technical detail needed by a software provider to support a function within a business process.
- The function might be something like 'Model a window'. This has various data requirements that need to be supported. A functional part includes the detailed information about those IFC capabilities that need to be supported.
- Several functional parts can be brought together to provide an 'exchange requirement model'. It is this model that provides the technical specification necessary for IDM to work.

Making it local

- When an exchange requirement is built from the IFC standard, it is very general – it uses what the functional parts give it. It is however often the case that the information needed in a business process differs slightly from place to place.
- Often, this difference can be explained by the conventions which are applied to working at a given location. For instance, determining quantities in a construction project is done differently in the UK and Germany.
- The objects from which the quantities are obtained are the same but the interpretation of data is different. IDM deals with this by providing a 'business rules' component. This is simply a set of rules that can be applied to an exchange requirement to 'configure' the information output.
- In this way, the same exchange requirement can be used to satisfy industry needs in different places and provide acceptable results in both.

Providing BIM guidance

- An aim of IDM is that, as software applications include 'switches' that enable exchange requirements to be used, guidance will be provided by (or for) BIM applications to show the user how best to apply the requirement.
- By doing this, the aim of providing detailed BIM guidance to users for projects data will be served. A bit of an idealistic viewpoint? It's happening with the GSA space submissions in the US.

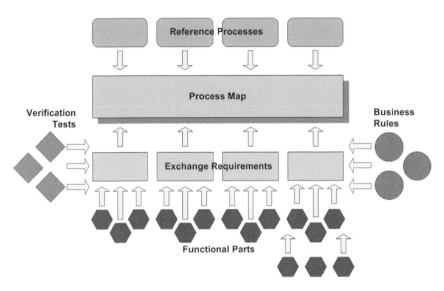

Figure 10.4 IDM technical architecture.

Supporting business processes

- A key aspect for making IDM relevant to the construction industry is that exchange requirements are identified through the provision of a process map.
- Process maps are developed using the 'Business Process Modeling Notation' (White 2005), which is an emerging standard from the Object Management Group (OMG 2006). It is an approach that has significant software tool support and which is widely used in other industries. Most significantly, it can also support migration of business processes to the Business Process Execution Language. This offers the potential for developing workflow control of IDM-developed exchange requirement models as a next stage in development.
- Each process map identifies the actors engaged in the business process in 'pools' (within which divisional functions might be identified in swimlanes).
- The information model is always identified as one particular actor in the process map. Communication with the information model is through the exchange requirements, which are data objects shown within the information model 'pool'.

Verifying data

- The final component of IDM is that the data provided through an exchange requirement can be automatically verified.

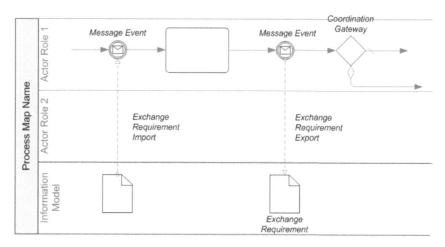

Figure 10.5 BPMN elaborations used in IDM.

Source: Wix (2007).

- Where the exchange requirement and the business rules say that certain data elements must be provided for defined objects and that their values must be within a given range, this can be verified without human intervention.
- The principle is exactly the same as for localizing an exchange requirement for checking building regulations. It is a simple comparison between the information needed and the information provided.

Working with certified software

Leading IFC-compliant software applications are certified through a testing and certification process operated by buildingSMART International. This relies on development of application 'views' that can be reliably supported by a software application. A 'view' is developed for use by the software implementers through the buildingSMART Model View Definition (MVD)

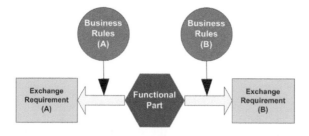

Figure 10.6 Applying business rules.

process (Hietanen 2006). Both IDM and MVD are now connected in an integrated IDM–MVD methodology that allows a complete specification of user requirements, technical solutions and project data exchange guidance.

The key connection between IDM and MVD occurs through the provision of the exchange requirement models. Each 'model view' provides a container for one or several 'exchange requirement models'.

Why do we want both exchange requirement models and model views?

- An exchange requirement model deals with information only at a single point in the process. It may be subject to business rules that localize or otherwise specify its usage and that may control how data populates the model. It can therefore be validated at a data level.
- A model view, however, deals with how a software application will work with IFC. It enables certification of the fact that software properly implements the entities (or classes) within the view but cannot determine how the data is populated. It will typically deal with multiple points in the process (e.g. an HVAC application software will deal with several HVAC design processes).

Figure 10.7 shows the typically expected flow of development through the integrated methodology. This is the illustration of the methodology included within the first issue of the North American National BIM Standard (Facilities Information Council of the National Institute of Building Sciences, USA 2007).

Conclusion

IDM together with Model View Definition (MVD) is beginning to address the issues of how the IFC model should be used appropriately for projects. It has come about because users had issues with the use of certified IFC applications, expecting that the certification process itself should resolve problems but not fully understanding that this was not the objective for certification. By focusing on user requirements and on the actual population of a building information model to support those requirements, IDM has introduced the idea of validation of data sets that is much more relevant to what users need to achieve in a project context.

IDM is a practical development. It came out of needs expressed by Norwegian users on hospital and college developments and is progressively supporting the needs of the State Building Agency (Statsbygg). It is being integrated with developments within European-funded projects and forms a key component in the delivery of the NBIMS standard (Gehre *et al.* 2006).

IDM in conjunction with MVD is set to deliver on the promise of process-driven data exchange. It does this by effectively creating a data exchange schema at every information exchange point but ensures that each such

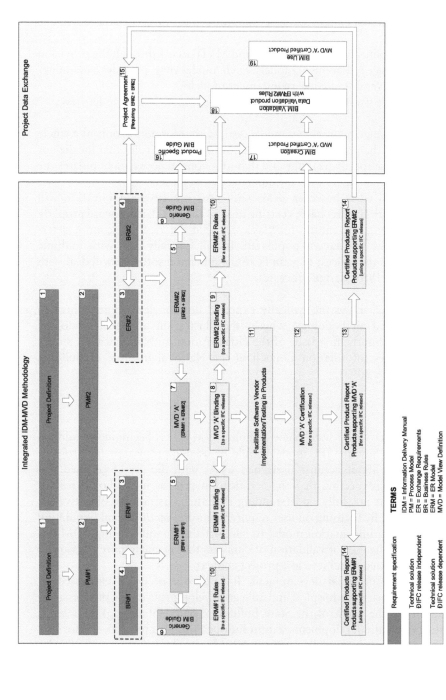

Figure 10.7 Overview of the integrated IDM–MVD methodology.

schema remains in conformance with the overall IFC model schema. It is this capability of delivering information at the micro-level whilst maintaining continuity at the macro-level that makes it such a powerful idea.

References

buildingSMART Norway (2007). *Information Delivery Manual: Guide to Components and Development Methods*, http://idm.buildingsmart.com

European Construction Technology Platform (2005). *FA Processes and ICT*, http://www.ectp.org/fa_pict.asp

Facilities Information Council of the National Institute of Building Sciences, USA (2007). *NBIMS Version 1.0 Part 1: Overview, Principles, and Methodologies*, http://nbimsdoc.opengeospatial.org/

Gallaher, M. P., O'Connor, A. C., Dettbarn, J. L., Jr and Gilday, L.T. (2004). *Cost Analysis of Inadequate Interoperability in the U.S. Capital Facilities Industry*, NIST GCR 04–867, USA, http://www.bfrl.nist.gov/oae/publications/gcrs/04867.pdf

Gehre, A., Katranuschkov, P., Wix, J. and Beetz, J. (2006). *InteliGrid Deliverable D31: Ontology Specification*. Ljubljana, Slovenia: InteliGrid Consortium, c/o University of Ljubljana, http://www.InteliGrid.com

Hietanen, J. (2006). *IFC Model View Definition Format*, International Alliance for Interoperability, April, http://www.iai-international.org/software/MVD_060424/IAI_IFCModelViewDefinitionFormat.pdf

International Alliance for Interoperability (2006). *IFC2x Edition 3, Industry Foundation Classes*, http://www.iai-international.org/Model/R2x3_final/index.htm

ISO (2005). *ISO/PAS 16739:2005, Industry Foundation Classes, Release 2x, Platform Specification (IFC2x Platform)*. Geneva: ISO.

Object Management Group (OMG) (2006). *Business Process Modeling Notation Specification*, OMG final adopted specification, http://www.bpmn.org/

White, S. A. (2005). *Introduction to BPMN*. New York: IBM, http://www.bpmn.org/Documents/Introduction%20to%20BPMN.pdf

Wix, J. (2007). *Quick Guide: Business Process Modeling Notation (BPMN)*. Oslo: buildingSMART Norway, http://idm.buildingsmart.com

11 Applying process rigour to the use of BIM in building design teams

A review of three technologies

Godfried Augenbroe

Abstract

Building design requires an orchestrated team effort in which many actors, tasks and activities have to be coordinated. As different actors use different software tools, each specialist traditionally operates on an island of isolation until the time comes to match and patch with other members of the design team. To accomplish this effectively one needs to be able to execute a wide variety of software applications in an efficient manner. This has led to the need for 'interoperability' between software applications. For the last 20 years, a sustained research effort has been devoted to achieving this in the architecture, engineering and construction (A/E/C) industry. This has led to the introduction of incrementally richer versions of a building information model (BIM) that integrates input and output data from different applications.

We argue that data integration is the cornerstone of true interoperability but that at least three additional technologies need to be applied. These technologies have to deal with coordination of the data exchange events, collaboration between the people using the software applications, and effective project planning of the deployment of a BIM solution. A personal account of three technology developments, one for each layer, will be given.

1 Introduction

What started in the 1960s and 1970s as one-to-one 'interfacing' of applications was soon realized to be non-scalable. In the 1980s therefore work started on the development of shared central building information models which would relieve the need for application-to-application interfaces, as depicted in Figure 11.1. The increased level of connectivity that could be achieved was termed 'interoperability', as indeed different applications would be able to 'interoperate' through the shared data model, at least in principle. The physical data exchange takes place through software interfaces that perform mappings between global (neutral view) and native (application view) representations (Eastman 1999).

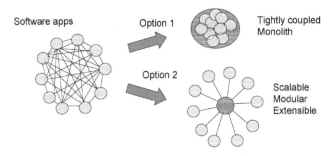

Figure 11.1 From non-scalable to scalable interoperability solutions.

It should be realized that the building industry has characteristics that make the development of a 'building product model' a huge undertaking. Successful applications are being deployed in other, highly organized engineering disciplines underpinned by systems for concurrent engineering, project data management, data sharing, integration and product knowledge management. In the building industry, full-blown systems have not reached the market yet. Major obstacles are the scale and diversity of the industry and the 'service nature' of the partnerships within it. The latter qualification is based on the observation that many relationships in a building project put less emphasis on predictable and mechanistic data collaboration than on the collaborative (and often unpredictable) synergy of human relationships. For this reason it is not a given that the industry is going to opt for open modular interoperability (option 2 in Figure 11.1) over tightly coupled application development (option 1). In this chapter we will concentrate only on the use of BIM in open architectures (option 2).

The average building project requires the management of complex data exchange scenarios with a wide variety of software applications. Building assessment scenarios typically contain simulation tasks that cannot be easily automated. They require skilled modelling and engineering judgement by their performers. In such cases only a certain level of interoperability can be exploited, stopping short of automation. This is the rule rather than the exception during the design phases, where designers call on domain experts to perform design assessments. Such settings are complex task environments where the outcome is highly reliant on self-organization of the humans in the system.

In projects, teams of experts are formed around specific design objectives. This requires coordination of tasks, timely exchange of information, and execution of 'local', discipline-specific software. But an effective project requires more than coordination of 'robotic' tasks; it requires a plan of collaboration between real people who are willing to co-plan their work, share knowledge and combine expertise to reach a common goal.

It is the purpose of this chapter to reflect on software interoperability in complex project settings, where data exchange between software applications cannot be separated from the project's coordination and collaboration layer and, ultimately, its planning. Adequate management of this group process must guarantee that design decisions are taken at the right moment with the participation of all involved disciplines. This must be prepared by adequate coordination of tasks and formation of a col-laborative project environment, and preceded by adequate planning of the project. It is argued that there are three layers – data exchange, coordin-ation/collaboration and planning – that need to be looked at in one common framework to arrive at interoperability solutions that work in real-life situations.

We will report on three technology developments in which the author was involved. Each technology development targets a specific layer, or a combination of them, identified above. All three developments apply their own flavour of process rigour to the use of BIM in complex project settings.

In this chapter we will focus on the integration of design assessment tools in design analysis settings where these issues play a dominant role. It will be argued that interoperability according to Figure 11.1 is a requirement but by no means the solution in highly interactive, partly unstructured and unpredictable design analysis settings. In those situations the support of the underlying human aspects of the designer-to-consultant interactions, as well as the special nature of their relationship, should be reflected in systems and support platforms. Different levels of integration among the team members of a building design team will have to be accomplished. First of all, there is the problem of the heterogeneity of information that is exchanged between one actor and another. Harmonizing the diversity of information in one common and consistent repository of data about the designed artefact is the first (traditional) level of ambition. The next level of integration is accom-plished in the total management, coordination and supervision over all communication that occurs within a project team. The third level must also reflect that the execution of project management must be grounded in a collaboratively developed project plan that should guarantee transparency, predictability and accountability across the project team.

It is hard to achieve interoperability of design tools in complex design settings. A fixed template of 'design analysis' interactions does not exist and cannot be assumed unless in predictable, static or pre-conditioned settings. In the idiosyncratic, spontaneous and self-organizing behaviour that is so common for building teams, these assumptions do not hold. With the trend towards dispersed teams of experts who can collaborate anywhere and any time, all offering their unique combination of expertise and tools, these systems have to offer open, flexible interoperability. An extended discussion can be found in Augenbroe and Eastman (1998), which is briefly summar-ized below:

- Current product models and standards are focused on data exchange; they do not take process context into account and therefore are unable to deal properly with data exchange control issues that are related to process logic.

- Current developments in building product models focus on single uniform ('neutral') building models. Yet neutral models have some distinct disadvantages. Firstly, interfaces between neutral models (containing all available data about a building) and specific tools (dealing with one performance aspect only) have to filter out only the relevant information, making these interfaces overly complex ('over-engineered'). Secondly, the mapping of data in typical design domains to technical and performance evaluation domains (e.g. to lighting or acoustics) might not be possible. In fact, current interoperability research has failed to address the fundamental issue of the computability of this type of cross-domain mappings. Thirdly, the use of neutral models might have implications for the order of execution of the steps in a building design process, imposing a rigid order for the use of tools and models.
- Current product models and standards assume that all information about a building design is well structured and stored in 'structured idealizations' of reality. Yet, as a fact of life, a vast proportion of information will remain to live only in unstructured media such as text documents, informal memos, personal notes and so on.
- Current product models assume that data mapping can be automated; this ignores that there will always be some need for additional expert-driven idealizations, based on schematization skills and engineering judgement.

In spite of the fact that the referenced white paper was written 10 years ago, many observations are still as relevant as they were then. The IFC development underlying current BIM manifestations has led to many working mappings, but fundamentally little has changed, and we argue that a thorough rethinking of a system approach to interoperability is still an urgent priority. We discuss three trials from the past decade that have proposed interesting approaches to achieve true interoperability on the three levels.

2 Coordination: the COMBINE Exchange Executive

Figure 11.2 shows the addition of a supervisory module around the central building model of Figure 11.1. It assumes that the building model is persistently stored in a database, obeying STEP development principles, as adopted in the COMBINE effort (Augenbroe 1995). The added supervisory component acts as an Exchange Executive (ExEx) that applies process logic to the occurrence of data exchange events (Amor *et al.* 1995).

The ExEx 'polices' all data exchange events and enforces the rules that govern the pre- and post-conditions for the execution of a data exchange event, that is, the execution of the input and output mappings between the native application view and the BIM view. These rules and conditions

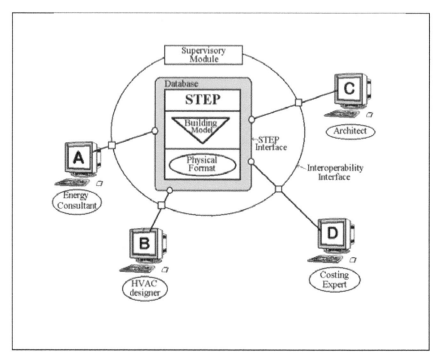

Figure 11.2 Process driven interoperability in COMBINE effort.

are developed in conjunction with the BIM. They express whether or not an application can be executed, for example by checking whether all necessary upstream applications have been called. The logic for this is to a great part derived from comparing the input and output schemata of all applications in the 'project window' of the project; in the case of Figure 11.2, the project window would contain the applications A, B, C and D.

The primary function of the added process management component was to manage all data transaction events. Two forms of control are regarded:

- *temporal control:* rules and conditions determining when a specific actor can perform a particular (simulation or design) operation;
- *data integrity controls:* rules and conditions to ensure that the BIM's instance data remains in a consistent state while the design progresses (Lockley and Augenbroe 2000).

The approach that was taken in the COMBINE project implemented the two forms of control in two separate system components. Data integrity control was implemented as 'constraint sets' in the database. The temporal control was based on the control logic embedded in a so-called 'project window' modelling formalism. This formalism was based on coloured Petri-Nets with conditions that could reason on the entities in the input and output schemata. The Petri-Nets contain information about all communicating 'actors' in the system (software applications and other tools) as well as their input and output schemata. In addition they contain a formal description of the order in which the actors are allowed or required to execute certain operations during design evolution, that is, obey logic dependency rules. The resulting event model formally defines the exchange event sequencing control. The Exchange Executive component then uses the event model to control the transitions, for example check constraints and inform an actor/application pair about the data availability and the ability to perform a simulation as the next step in the process.

At start-up of a project, the event model is read by the ExEx, which then automatically self-configures to assist the project manager at runtime to decide which application can be executed next and which applications are affected by changes to the building model and therefore need to be rerun. The state of the project moves forward as tokens move through the network of 'places' and 'transitions' on the screen. It allows a team manager to decide whether a software can be deployed next, based on the automatic checking of constraints by the ExEx. The functioning of the ExEx was demonstrated on two 'virtual design offices'. It allowed the project manager to control and manage data exchange events and avoid deadlocks and pre-emptively avoid application interfaces crashing because expected BIM object classes had not been instantiated yet. In such cases the ExEx gave enough information to the project manager to put the project back on track.

The ExEx was the first application of its kind to bring a level of data exchange coordination to the use of BIM. The approach is elegant in that it combines data dependency rules with application sequencing logic. It also recognizes that the human-in-the-loop approach to project coordination is the preferred method to make interoperability work. It is surprising that the ExEx approach has not received more attention in ensuing attempts to bring coordination to interoperability.

3 Collaboration: the design analysis integration (DAI) workbench

Design analysis integration deals with the effective use of existing and emerging building performance analysis tools in design analysis scenarios, with the participation of a team of designers and consultants. Some of the longer-term objectives are better functional embedding of analysis tools in the design process, increased quality control for building analysis efforts, and exploitation of the opportunities provided by the Internet. The latter refers to the possibilities for collaboration in loosely coupled teams where the execution of specific building performance analysis tasks is delegated to (remote) domain experts. It is obvious that in such teams, process coordination and collaboration are the critical factors, whereas data inter-operability is a 'support act' rather than the main objective.

Design analysis is performed through the complex interplay between design activities and analysis efforts by experts with an arsenal of simulation tools, testing procedures, expert skills, judgement and experience. Different paradigms of expert intervention in the design processes are described in Chen (2003). The scope of our treatment of design analysis integration is limited to the assumption that the design team generates specific design analysis requests, leading to an invocation of the input of (a team of) analysis experts (Figure 11.3), engaging in a design analysis dialogue. Figure 11.3 shows the complexity of the dialogue, and it is clear that the interaction between the design and analysis actors goes well beyond any current notion of tool interoperability.

Multiple information sources describe the state of the design at the time of the analysis request. This information is contained in different structured and unstructured documents, for example in unstructured documents such as drawings, specifications and so on, in semi-structured documents such as CAD files, and also partly in highly structured documents such as populated IFC models. Within a design decision, an expert may be consulted upon which a design analysis request is generated in some formal manner (accompanied by a contractual agreement). Upon fulfilment of the request and completion of the analysis, an analysis report is submitted to the design team. This report is then used to perform a multi-aspect evaluation by the design team in order to fully inform a pending design decision. This evaluation may lead to the generation of new design variants for which a follow-up analysis request is issued. In many instances, a comparison of

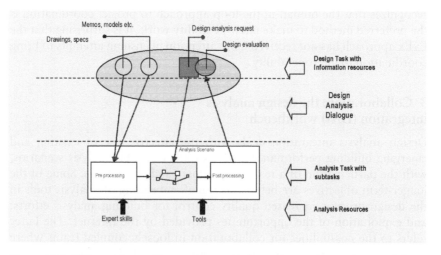

Figure 11.3 Different types of interaction and information exchange that result from a design analysis request.

design variants may already be part of the original design request and thus part of the submitted analysis report.

As Figure 11.3 suggests, there are multiple interactions and information flows between design and analysis tasks, each of them constituting a specific element of the dialogue that needs to take place between the design team and the analysis expert. In past work, the main emphasis has been on support for the data connections between the design representations and the input or native models of the simulation tools. Little work has been done on the back end of the analysis. Back-end integration requires the formal capture of the analysis results before they are handed back to the design team. The most relevant work in both areas is linked to the use of the IFC as structured design representation and recent work on representation of analysis results embedded in the IFC (Hitchcock *et al.* 1999).

A design analysis framework should cover all elements of the interaction and their implementation in a configurable communication layer that drives the interoperable toolset. The next sections explain the prototype that was developed to demonstrate this.

3.1 A workbench for design analysis interactions

The workbench encapsulates data interoperability in order to capitalize on efforts that have been invested in the development of building product models over the last decade. It should not however be based on any limiting assumptions about the design process or the logic of the design analysis interaction flow. The framework should offer support for the interaction between the building design process and a wide array of building performance analysis tools. This can be realized through a 'workbench' with four layers. The workbench positions building design information on the top layer and simulation applications (and more generically 'analysis tools') on the bottom layers. In order to move from design information to analysis tool (pre-processing) or from analysis tool to design-relevant information (post-processing) one has to pass through two intermediate layers. Those intermediate layers provide context to a specific interaction moment by capturing information about the process and information about the structure of the exchanged data on two separate layers, as shown in Figure 11.4. The two intermediate layers are the key layers of the workbench. They allow the domain expert to manage the dialogue between the design team (top layer) and the analysis applications (bottom layer).

Interoperability typically makes a direct connection between the top and bottom layer through a data-mapping interface. The four-layered workbench is the 'fat' version of this traditional view on interoperability. The top layer contains all building design information in partly structured, partly unstructured format. The building model layer contains semantic product models of varying granularity that can be used for specific analysis and engineering domains or specific performance aspects. The scenario layer

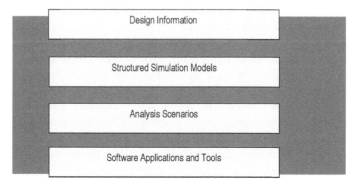

Figure 11.4 Four-layered workbench to support design analysis dialogues.

captures the process logic (workflow), allowing the planning of a process as well as the actual 'enacting' of that process. These functions are offered by current mainstream workflow design and workflow enactment applications. The bottom layer contains software applications (mainly building performance simulation tools) that can be accessed from the scenario layer to perform a specific analysis. Analysis functions, rather than specific software applications, are activated, removing the dependency of the workbench on particular simulation software packages. This concept is fundamental to the workbench. It is based on the introduction of a set of predefined 'analysis functions'. Analysis functions act as the smallest functional simulation steps in the definition of analysis scenarios. Each analysis function is defined for a specific performance aspect, a specific building (sub)system and a specific measure of performance. An analysis function acts as a scoping mechanism for the information exchange between design information layer, model layer and application layer.

The following is fundamental to the intended use of the workbench:

- The workbench is process-centric, which allows for explicit definition, management and execution of analysis scenarios. These scenarios will typically be configured by the project manager at start-up of a new consulting job. The fact that a job can be explicitly managed and recorded offers an additional set of functions for the architectural and engineering office. Audit trails of a building analysis job can be stored, and previous scenarios can be reused in new projects. This can potentially provide a learning instrument for novices in simulation. In-house office procedures and scenarios can be stored for better quality assurance.
- Expert knowledge and expertise are essential elements of performance assessment. Judgement of the applicability of performance assessment methods and evaluation of the validity of results obtained with (computational) tools are essential human skills that the workbench recognizes.

3.2 The central role of the scenario layer

The workbench is deployed when a design interaction moment occurs and a design analysis request is triggered. The analysis expert will use the scenario layer to define the task logic of the analysis steps that respond to the request. As an example, the situation can be envisioned where an architect contacts an expert consultant in the selection of a glazing system in an office space. The expert consultant discusses the actual analysis task with the architect, plans the steps needed to carry out the analysis, and assigns one of the in-house simulation experts to carry out one or more of the analysis tasks in the scenario, as depicted in Figure 11.5.

The scenario layer typically offers access to a commercial workflow process modelling tool to define the analysis scenarios. This type of software offers a graphical front end to a variety of 'workflow enactment engines', that is, computer programs that automate the dispatching of tasks to assigned task performers, transfer of documents, coordination of dependencies between information, tasks, tools and actors in an organization, and so on. It allows the planning of a scenario that covers all steps of the analysis process, that is, from the initial design analysis request issued by the design team to the close-out of the consultancy job. It details the actual execution of the analysis, anticipates potential mid-stream modification of the analysis plan, and plans the feedback that is to be provided to the designer/architect. It allows graphical representation of tasks in process flow diagrams that can be constructed using drag-and-drop capabilities. It also allows easy decomposition of complex processes. A screenshot of the process modelling window is shown in Figure 11.6, showing one of the tools available for this purpose (Cichocki *et al.* 1998).

One of the expected advantages of using a generic workflow engine is the easy integration of the workbench in environments where workflow management is already used to manage businesses processes. The integration of the simulation process with internal business processes such as invoicing, reporting and resource allocation within the same firm or across collaborating firms is an exciting future prospect for the DAI workbench. It would indeed add significantly to project management and quality assurance and on-the-job training within the engineering enterprise.

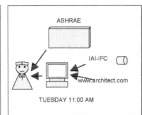

Figure 11.5 Start of the analysis process.

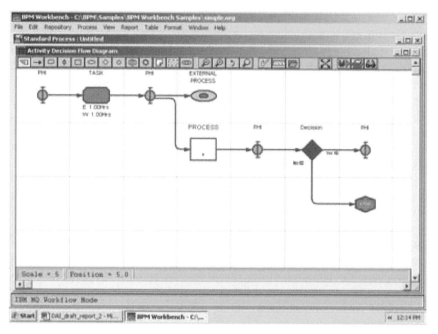

Figure 11.6 Workflow modelling window for analysis scenarios.

3.3 *The analysis function concept*

Analysis functions are the key to the connection of the scenario layer with the building simulation model on the one hand and the software application on the other. They allow the expert to specify exactly what needs to be analysed and what results (captured as quantified performance indicators) need to be conveyed. To do so, analysis functions need to capture the smallest analysis tasks that routinely occur in analysis scenarios. Each analysis function (AF) must identify a well-defined (virtual) experiment on the object, which is defined to reveal building behaviour that is relevant to the performance aspect that is to be analysed. The AF must be defined in a tool-independent way, and formally specified by way of an AF schema. The AF schema defines the data model of the building system that the AF operates on, as well as the experiment and the aggregation of behavioural output data. The analysis itself is fully embodied in the choice of an AF. However, not all AF calls need to be performed by a software application. For instance, an AF may define the daylighting performance of a window as the result of physical experiment, for example by putting a scale model in a daylight chamber. Other AFs may be defined such that the measure is qualitative and subjective, based on expert engineering judgement. Allowing all these different types of AFs to be part of the same environment and controlled by a transparent workflow model adds to the control over the process, especially if one realizes that many different experts may be called upon in the same project, each using their own analysis expertise.

Note that this approach is conceptual and disconnects the AF completely from its 'incidental' software realization. For each individual AF, the entities and attributes that are described by the experiment are based on generalizations of performance analysis.

3.4 The DAI workbench prototype

The framework introduced in the previous sections has been implemented in the Design Analysis Interface-Initiative (Augenbroe and de Wilde 2003; Augenbroe *et al.* 2004). Its main objective was to test the proposed framework with focus on the scenario development and the embodiment of AF-driven interfaces to existing simulation tools.

At runtime an AF model needs to be populated through data interfaces that 'pull' data from the models that reside in the building analysis model layer. The 'pull approach' allows for a directed search of relevant information that resides on the upper layers of the workbench. Whenever building information is present in structured format, the data exchange from the upper layer to the AF model can be automated. However, the AF model population interface may signal missing information and trigger manual input by the simulation expert. Since the AF model describes only the elements that are critical for a specific performance analysis, both the automatic part of the interface and the user-guided constructive part represent small and manageable programming efforts for the developers. They will benefit greatly from emerging XML-based graphical (declarative) mapping languages. Each AF drives the connection and data exchange with the neighbouring workbench layers. The principle of the approach is shown in Figure 11.7.

Only a rudimentary test of the workbench could be performed with three AFs, that is, for energy efficiency, thermal comfort and daylight autonomy.

3.5 Evaluation

The eventual success of the DAI approach to building a coordination and collaboration platform for the robust execution of interoperability will depend on the following issues:

- The runtime, user-driven population of subschema information requires a new generation of development tools for 'constructive interfaces'. If a task in the workflow (layer 3) points to one of the predefined subschemata (on layer 2), a mapping task is activated to map BIM data into the subschema. This mapping is however not automatic, and needs intervention by the user to add missing data and perform steps to

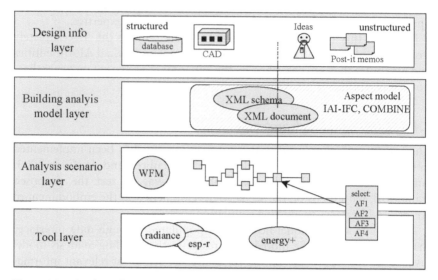

Figure 11.7 Analysis function as central driver in the data exchange topology.

generate a 'right-engineered' model, all the while supported by rich access to the unstructured design information (layer 1).

- The concept of software-independent definition of atomic AFs is not an absolute requirement for the DAI approach, but it would give the workbench increased flexibility and independence from specific software applications.

Both issues need more study and a larger-scale trial to check their feasibility.

4 Planning: a collaborative planning platform based on structured dialogues

Project planning in the construction industry requires a lot of experience from individual planners. Existing planning software tools typically provide many features to list and organize design and construction activities, to optimize resource allocations, to visualize expected project schedules, to coordinate project execution, and so forth. But, in order to arrive at a meaningful project plan in the first place, users must possess project- and domain-specific know-how, with anticipation and improvisation skills that are not provided or supported explicitly by the offered planning environment. Even the user's complete familiarity with the software functionality provides no guarantee for the quality of a generated project plan and whether its content is actually sensible, reliable and comprehensive. Moreover, most planning tools are not collaborative in the sense that they do not specify who is responsible for generating what planning data, and when.

Instead it is often a single project manager who assimilates the project plan. A lot of embedded planning knowledge is thus unscripted, undocumented and preserved only in personal memories or in scattered documents, making its transfer and enrichment over time limited to interpersonal contacts and people's direct project involvement.

In an attempt to increase predictability and efficiency, and to improve knowledge retention across projects, a more systematic approach to project planning was proposed and tested in a prototype.

We argue that the same approach has great potential for the planning of BIM deployment.

The approach is based on the notion of a meta-process model for planning activities. It embodies and cultivates the logic and intelligence of incremental and collaborative planning activities in a given domain. Planning tasks are encoded and enforced as a set of structured dialogues between project partners. Two applications have been tested thus far: 1) systematic planning of ad hoc e-engineering partnerships (Augenbroe 2004; Ren *et al.* 2006); and 2) the application in owner–architect planning situations (Verheij and Augenbroe 2006). Both prototypical applications were built on top of an existing web-based collaborative virtual environment (CVE). The platform supports the enactment of workflows that are instances of the process planning meta-model, and combines this with the support of structured dialogues.

A more methodical approach to project planning identifies the procedural steps for project partners to systematically work towards a solid project plan. The applied mechanism to realize process-driven project planning is the execution of a planning workflow. However, traditionally workflows tend to be rather mechanistic, dispatching individual rather than collaborative activities. Therefore the concept of workflows is expanded to include the notion of so-called 'structured dialogues'. A structured dialogue not only assigns consecutive tasks according to a predefined logic but is designed also to support the collaborative nature of planning activities.

The basic approach is driven by an 'ushering' and dialogue capturing metaphor (Figure 11.8). At different stages of the planning effort, different combinations of project partners are ushered into a virtual meeting room, whenever needed. Once inside the meeting room, their interaction is guided by the meta-planning process model. This model structures and (at runtime) enacts the interaction between planners in such a way that the dialogue is productive and efficient. The steps in the dialogue can be ad hoc and informal, and they can be supported by multiple channels (e.g. instant messaging and phone calls), but the outcome needs to be hand-reported in named fields on the whiteboards that line the virtual meeting room.

BIM relevance. There is good reason to believe that the adoption of BIM in a project at start-up would be greatly enhanced by such a mediated project planning effort.

The objective is to have groups of project planners, for example

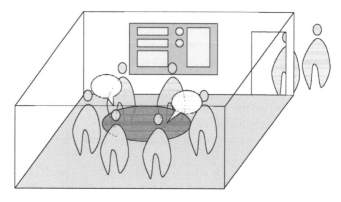

Ushering project partners in a meeting room
with whiteboards containing data fields to fill

Figure 11.8 The ushering paradigm.

companies or project teams, execute workflows in which the system prompts participants for each activity to access, submit or edit BIM-related deliverables that were generated upstream or that are needed downstream. The data templates ('shared whiteboards') that are thus incrementally populated during the virtual planning meetings are together intended to form a comprehensive project plan that addresses issues like risks, rewards, responsibilities, arbitration, deadlines and quality assurance related to the planned use of BIM in a project.

The technology. Workflows are expressed in the neutral XML Process Definition Language (XPDL), a standard put forward by the Workflow Management Coalition (WfMC) in an effort to stimulate interoperability between workflow software vendors or execution engines.

A prototype was built on the substrate of an existing CVE developed in the European e-HUBs project (Augenbroe 2004), which focused on the brokering between clients and providers of e-engineering services. This prototype provides an online collaboration platform with built-in communication and community and document management, but it also comes with a customized process modeller and an embedded workflow enactment engine. Figure 11.9 shows an example of a planning meta-model, in this case for the design build selection case, developed in the Java Workflow Engine (JAWE) tool.

The process model contains 'swimlanes' for each potential actor in the planning process. Each task is linked to other tasks through logic relationships, and a common set of 'whiteboards' (attribute fields), that get instantiated during the execution of the tasks. Each attribute set is linked to a subset of tasks and swimlane/actor. Every interaction with an attribute is governed by access right (read, write or read/write). This corresponds to the metaphor of planners being in the same room, having pens, and having access to

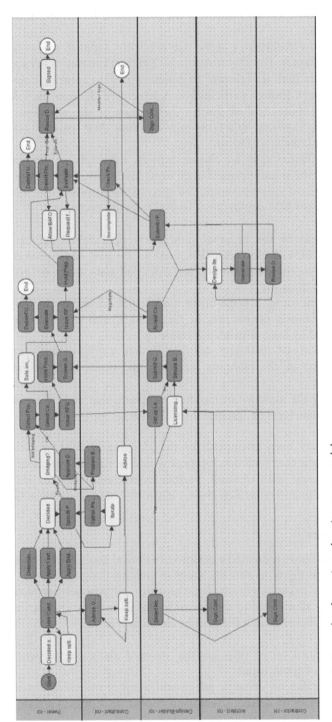

Figure 11.9 Example of a project planning meta-model.

different field on the whiteboards in the plan room for either viewing or writing or both.

The workflow technology does not explicitly support the notion of structured dialogues. For this reason, the XPDL taxonomy was expanded to cover aspects that are relevant to conducting and capturing controlled meetings between project participants. A structured dialogue is linked to each set of attributes (called 'form') that is part of the workflow. Each dialogue type is captured by a taxonomy of terms, their semantic relationships and their linkage to a form.

Evaluation. Planning is an elusive activity, especially in construction projects where the situation can change from day to day. Yet many projects fail because of inadequate planning. To execute the proposed technology in the preparation of BIM deployment, several steps are necessary:

- identification of the minimal (and fairly predictable) set of issues regarding BIM deployment that need to be part of a project plan (quality assurance, constraints, rewards, skill tests, software versions, legal obligations, etc.);
- development of a process meta-model that identifies the different roles of planners (swimlanes in Figure 11.9) and their interactions, constrained by logic dependency rules;
- identification of dialogue attributes that capture the outcomes (resolutions) of the planning activities, the result being a dialogue taxonomy;
- structuring the dialogue in interface templates that drive the planners to reach some form of agreement on all issues.

This has never been attempted for the planning of BIM in a real-life project. We argue that such development would be valuable in understanding BIM planning with multiple partner organizations having different agendas and with different roles, skills, obligations and reward expectations.

5 Conclusions

In this chapter I have presented a personal, historical account of three efforts to apply process rigour to data exchange in building design projects.

Although in many ways similar, the three approaches are distinctly different. They all respond to the typical setting in architectural/engineering design offices (Figure 11.10) that can be characterized as:

- No predefined set of rules can be assumed about the sequence and operation of applications.
- The team relies on dynamic self-organization of individual actors; the need for flexibility defeats attempts to impose strict project management rules or assume pre-conditioned project settings.

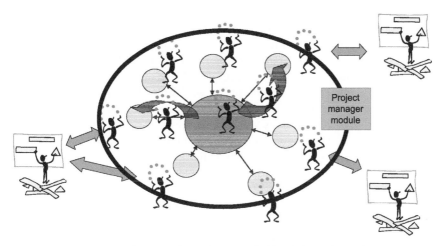

Figure 11.10 The use of BIM in the architectural office follows few rules.

- The project environment cannot be viewed in isolation from the firm environment.
- There is poor visibility of responsibilities, ownership and accountability with respect to collaborative tasks.

This situation can hurt the effective deployment of BIM.

We have provided an account of three approaches that were developed over the last decade to respond to this situation at different levels:

- The ExEx attempts to provide a necessary level of execution management to the use of BIM. It relies heavily on the explicit definition of input and output schemata for each application. It infers from that information when it is possible or necessary to run a certain application. It can be said that 'open' interoperability solutions (i.e. where no pre-conditioned process can be assumed) cannot function without a component that regulates and manages the execution sequence. In spite of the great strides in BIM development such a component is still missing. Indeed most successful BIM solutions work in controlled and predefined scenarios. For architectural design situations the lack of flexible yet secure control of data exchange events is a major obstacle for wider BIM adoption.
- The DAI approach does not rely on inferred logic for sequencing different data exchange events. It puts all control back in the hands of the project manager, relying on rapid definition of execution scenarios using workflow management technology. In doing so the project manager can respond to potential frictions between project and office environments, as it is easy to combine project-specific demands with firm-specific circumstances. This is novel, as it broadens the environment

in which the use of the BIM can be controlled. It adds in fact the opportunity to integrate the very data management aspects in the control layer. It takes a radical departure from the subset approach in COMBINE by asserting that it is not efficient to look at applications as one functional entity with one input and one output schema. Rather, it regards applications as having a set of distinct functions (not from an internal software function perspective, but from an external design analysis perspective). This way one arrives at a distinct set of 'analysis functions'. This set can then reside on the function pallet of the workflow designer, who can use them as designated tasks in the design of a workflow for a specific project. Moreover, for every design AF there is an identified input and output schema, and an application interface can be pre-made for that schema. However, what seems to be the big advantage of the approach is also its biggest challenge: it could prove very hard to come up with a manageable set of (parameterized) AFs.

- The third approach takes an even broader view on the deployment of BIM; it addresses the need for adequate project planning to address issues such as QA, accountability, responsibilities, rewards, and other aspects that form part of a partner agreement. Those are crucially important issues when different legal entities collaborate in a project. It was shown that with the right mix of collaboration technologies, and predefinition of a planning model and attribute sets of the resulting project plan, a BIM planning approach could be constructed.

The three efforts reflect the research journey made by the author over the last 15 years in his quest to enable the deployment of BIM in complex design settings. No attempt has been made to give an account of similar efforts deployed by other researchers in the same period. Instead, the developments have been discussed from a common viewpoint to support a comprehensive view of the lessons that were learned and the road towards BIM deployment that stretches out before us. One common thread, in the development track, albeit more implicit than intentional, is that the author's research focus has shifted over time. Whereas the earlier work had an 80 per cent technology and 20 per cent human focus, the latest efforts have shifted towards a 20 per cent technology and 80 per cent human focus. This confirms the lesson that an old and wise teacher conveyed to the author when he was a student. But it seems that this lesson could be learned only by experiencing the limitations of one's technology-driven solutions first-hand.

References

Amor, R., Augenbroe, G., Hosking, J., Rombouts, W. and Grundy, J. (1995). 'Directions in Modeling Environments', *Automation in Construction*, 4, pp. 173–187.

Augenbroe, G. (1995). *COMBINE 2: Final Report*. Brussels, Belgium: Commission of the European Communities, http://dcom.arch.gatech.edu/bt/Combine/my_www/document.htm

Augenbroe, Godfried (2004). 'e-HUBs: e-Engineering Enabled by Holonomic and Universal Broker Services', in P. Cunningham and M. Cunningham (eds), *Proceedings eChallenges, eAdoption and the Knowledge Economy: Issues, Applications, Case Studies*, pp. 1285–1292. Vienna, 27–29 October 2004. Amsterdam: IOS Press.

Augenbroe, G. and Eastman, C. (1998). 'Needed Progress in Building Design Product Models', White paper, Georgia Institute of Technology, Atlanta, USA.

Augenbroe, G. and de Wilde, P. (2003). *Design Analysis Interface (DAI): Final Report*. Atlanta: Georgia Institute of Technology, http://dcom.arch.gatech.edu/dai/

Augenbroe, G., Malkawi, A. and de Wilde, P. (2004). 'A Workbench for Structured Design Analysis Dialogues', *Journal of Architectural and Planning Research*, 21(4), pp. 321–330.

Chen, N. Y. (2003). 'Approaches to Design Collaboration Research', *Automation in Construction*, 12(6), pp. 715–723.

Cichocki, A., Helal, A., Rusinkiewicz, M. and Woelk, D. (1998). *Workflow and Process Automation, Concepts and Technology*. Amsterdam: Kluwer.

Eastman, C. M. (1999). *Building Product Models: Computer Environments Supporting Design and Construction*. Boca Raton, FL: CRC Press.

Hitchcock, R. J., Piette, M. A. and Selkowitz, S. E. (1999). 'A Building Life-Cycle Information System for Tracking Building Performance Metrics', Proceedings of the 8th International Conference on Durability of Building Materials and Components, 30 May – 3 June 1999, Vancouver, BC.

Lockley, S. and Augenbroe, G. (2000). 'Data Integration with Partial Exchange', *International Journal of Construction Information Technology*, 6(2), pp. 47–58.

Ren, Z., Anumba, C. J., Hassan, T. M., Augenbroe, G. and Mangini, M. (2006). 'Collaborative Project Planning: A Case Study of Seismic Risk Analysis', *Computers in Industry*, 57(3), April, pp. 218–230.

Verheij, J. M. and Augenbroe, G. L. M. (2006). 'Collaborative Planning of AEC Projects and Partnerships', *Automation in Construction*, 15, pp. 428–437.

12 Building information modelling in material take-off in a Hong Kong project

Kenny T. C. Tse, Andy K. D. Wong and Francis K. W. Wong

Abstract

Building information modelling (BIM) is an emerging technology in the construction industry, although the concept has existed for more than two decades. With more BIM pilot projects in place, the industry has an increasing interest in and body of BIM knowledge. The experience and findings using BIM in materials take-off in a Hong Kong project are presented in this chapter. A common current contention is that BIM could replace quantity surveyors in construction projects. To examine the possible veracity of such a statement a study has been made of the differences between model quantities and bills of quantities according to the stipulated standard methods of measurement. The examples and illustrations are taken from a project in which the authors are involved, together with an intensive literature review. The findings clearly show the limitations of BIM in materials take-off and the importance of quantity surveyors in that process. Quantity surveyors were observed to have a strong involvement with BIM in the project design process, stronger and more pertinent than those involved in the traditional two-dimensional drafting practice.

Keywords

Building information modelling, BIM, taking off, bills of quantities, Hong Kong.

1 Introduction

Building information modelling is defined as 'a data-rich, object-based, intelligent and parametric digital representation of the (building) facility' (GSA 2006). Unlike traditional two-dimensional computer-aided design (CAD), all the views/schedules in a building project are generated from a single model instead of separate and independent CAD files. As such, the views/schedules are always coordinated, consistent and correlate, whenever an action is taken in a building information model. Different software vendors describe BIM in slightly different ways and in accordance with their views or means of identification of the key procedures, despite the fact that their BIM software has similar characteristics (Autodesk 2007; Bentley 2007; Gehry Technologies 2007; Graphisoft 2007; Nemetschek 2007). In recent years, BIM has received increasing attention in the construction industry all over the world. An increasing uptake of BIM in large-scale real projects, such as Freedom Tower in the United States (Day 2005), Eureka Tower in Australia (Graphisoft 2007), One Island East in Hong Kong (Tse *et al.* 2006) and Pan Peninsula in the United Kingdom (Autodesk 2007), has been very much in the eye of the construction community, thus giving BIM world exposure. BIM or Digital Architecture has recently become a key topic at architectural engineering and construction (AEC) conferences and workshops (AIAB 2006, CAADRIA 2007, CIB-W78 2007, ETS-IV 2007, HKIA 2007, IAI-UK 2007, IAI-US 2007, I-ESA 2007). Special issues on BIM have been called for by international refereed journals (*ITCon* 2007, *Automation in Construction* 2007 and *Construction Innovation* 2007). In short the stage seems to be set for BIM to be in a position rapidly to revolutionize the construction industry. Thus the sharing of BIM experience among practitioners and researchers could be both energizing and practical.

2 Objectives and methodology

The objectives of this paper are 1) to disseminate a BIM case study and in particular 2) to evaluate the use of BIM in materials take-off. Two of the authors of this chapter acted as consultants for the project management, BIM and materials take-off. Thus the information presented is first-hand and without prejudice and comes with the authors' intention of highlighting positive experience and lessons learned.

3 The project

The project is a two-storey indoor swimming pool which is an extension to an existing secondary school in Hong Kong. It is situated on top of a cut slope where the bed rock profile varies from 3.9 metres to 26.8 metres below the existing ground level. The foundation comprises 76 mini-piles, with each pile penetrating 3.3 metres below the rock surface (Figure 12.1). The filtration plant room is integrated with part of the substructure. The tie-beams are arranged to support the bearing structure of the swimming pool cube (Figure 12.2). The ground floor houses a six-lane pool measuring 15 metres by 50 metres, a male and a female changing room, an external basketball court and associated performance stage and planters (Figure 12.3). The upper floor is designated as a store. Another basketball court on the roof is accessible through an external staircase and a walkway connecting the existing school (Figure 12.4). The pool is equipped with some 'green' features, including solar chimneys, reflector panels and photovoltaic panels. The

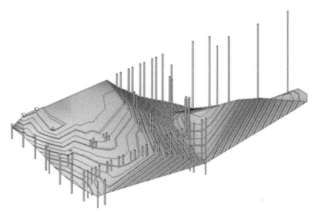

Figure 12.1 Rock profile and mini-piles.

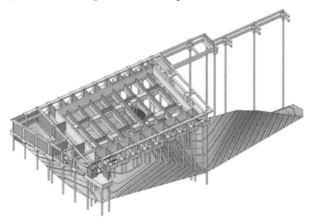

Figure 12.2 Pile caps, tie-beams and bearing walls.

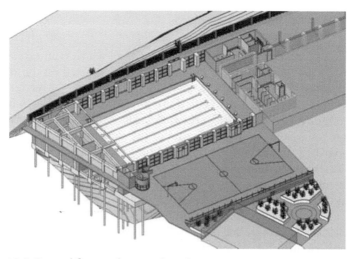

Figure 12.3 Ground floor and external works.

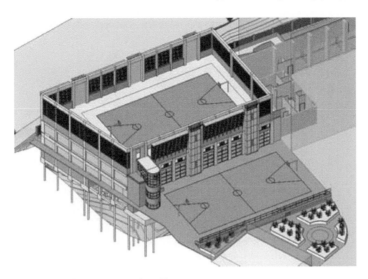

Figure 12.4 Basketball court and walkway.

design was awarded a 'Merit in Environmental Design' in the research and planning category by the Professional Green Building Council (PGBC 2006). The construction commenced in November 2006. The implementation of BIM in this project is led by the project manager (PM) and a BIM manager rather than the design consultants. The full details of the implementation framework are given by Tse *et al.* (2007).

4 The built-in materials take-off function

The integrated architecture and structure model was built using Revit Building 9.0 (Figure 12.5). Like most other BIM tools, this software package has a built-in materials take-off function. As a materials schedule is regarded as a model view, this function is found under the new view menu. Once the function is chosen, a new materials take-off dialogue box appears (Figure 12.6). The whole model is composed of different categories of building elements such as columns, windows and furniture. A single category or multi-category can be selected from given lists. The taking off can be targeted at either existing or new phases of construction. After making the selection, the materials take-off properties dialogue box pops up (Figure 12.7). The fields of the materials schedule can then be chosen from a list of available fields. The schedule content can be filtered, sorted, and grouped according to need. Once all the selections are made, the materials schedule is automatically tabulated (Figure 12.8).

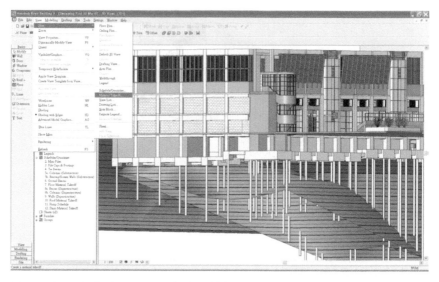

Figure 12.5 The built-in material take-off function in the BIM software.

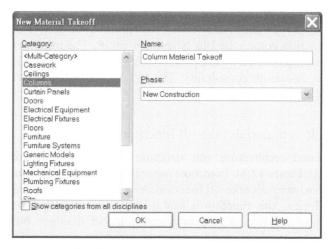

Figure 12.6 Material take-off category.

5 Model quantities versus bills of quantities (BQ)

A vitally important question is whether the quantities in the materials schedules (referred to as 'model quantity' in the rest of the chapter) can be used in contract bills of quantities. This leads to the understanding of how the model quantities are derived by the software. The model quantities of a number of objects were verified against the geometries. Table 12.1 shows the model areas and volumes of a concrete wall, structural column and

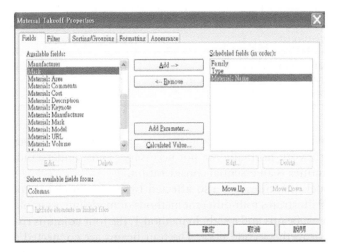

Figure 12.7 Material take-off properties.

Family	Type	Material: Name	Mark	Length	Base Level	Top Level	Top Offset	Material: Area	Concrete: Volume
R.C. Column (Superstructure)	400 x 850	R.C. Grade 40/20 (Superstructure)	C24G	2550	G/F	U/G Floor	-150	7.055 m²	0.87 m³
R.C. Column (Superstructure)	400 x 500	R.C. Grade 40/20 (Superstructure)	C35G	2700	G/F	U/G Floor	0	5.260 m²	0.54 m³
R.C. Column (Superstructure)	400 x 500	R.C. Grade 40/20 (Superstructure)	C37G	2700	G/F	U/G Floor	0	5.260 m²	0.54 m³
R.C. Column (Superstructure)	400 x 500	R.C. Grade 40/20 (Superstructure)	C25G	2700	G/F	U/G Floor	0	5.260 m²	0.54 m³
R.C. Column (Superstructure)	400 x 500	R.C. Grade 40/20 (Superstructure)	C36G	2700	G/F	U/G Floor	0	5.260 m²	0.54 m³
R.C. Column (Superstructure)	400 x 500	R.C. Grade 40/20 (Superstructure)	C26G	2700	G/F	U/G Floor	0	5.260 m²	0.54 m³
R.C. Column (Superstructure)	600 x 600	R.C. Grade 40/20 (Superstructure)	C22G	5350	G/F	R/F	-150	13.560 m²	1.93 m³
R.C. Column (Superstructure)	500 x 500	R.C. Grade 40/20 (Superstructure)	C34G	2550	G/F	U/G Floor	-150	5.600 m²	0.64 m³
R.C. Column (Superstructure)	500 x 500	R.C. Grade 40/20 (Superstructure)	C33G	2550	G/F	U/G Floor	-150	5.600 m²	0.64 m³
R.C. Column (Superstructure)	800 x 600	R.C. Grade 40/20 (Superstructure)	C21G	2550	G/F	U/G Floor	-150	8.100 m²	1.22 m³
R.C. Column (Superstructure)	800 x 600	R.C. Grade 40/20 (Superstructure)	C23G	2550	G/F	U/G Floor	-150	8.100 m²	1.22 m³
R.C. Column (Superstructure)	1150 x 600	R.C. Grade 40/20 (Superstructure)	C17G	5350	G/F	R/F	-150	20.105 m²	3.69 m³
R.C. Column (Superstructure)	1150 x 600	R.C. Grade 40/20 (Superstructure)	C19G	5350	G/F	R/F	-150	20.105 m²	3.69 m³
R.C. Column (Superstructure)	800 x 600	R.C. Grade 40/20 (Superstructure)	C6G	5350	G/F	R/F	-150	15.940 m²	2.57 m³
R.C. Column (Superstructure)	800 x 600	R.C. Grade 40/20 (Superstructure)	C5G	5350	G/F	R/F	-150	15.940 m²	2.57 m³
R.C. Column (Superstructure)	800 x 600	R.C. Grade 40/20 (Superstructure)	C7G	5350	G/F	R/F	-150	15.940 m²	2.57 m³
R.C. Column (Superstructure)	800 x 600	R.C. Grade 40/20 (Superstructure)	C9G	5350	G/F	R/F	-150	15.940 m²	2.57 m³
R.C. Column (Superstructure)	800 x 600	R.C. Grade 40/20 (Superstructure)	C11G	5350	G/F	R/F	-150	15.940 m²	2.57 m³
R.C. Column (Superstructure)	800 x 600	R.C. Grade 40/20 (Superstructure)	C13G	5350	G/F	R/F	-150	15.940 m²	2.57 m³
R.C. Column (Superstructure)	800 x 600	R.C. Grade 40/20 (Superstructure)	C15G	5350	G/F	R/F	-150	15.940 m²	2.57 m³
R.C. Column (Superstructure)	800 x 600	R.C. Grade 40/20 (Superstructure)	C8G	5350	G/F	R/F	-150	15.940 m²	2.57 m³
R.C. Column (Superstructure)	800 x 600	R.C. Grade 40/20 (Superstructure)	C10G	5350	G/F	R/F	-150	15.940 m²	2.57 m³
R.C. Column (Superstructure)	800 x 600	R.C. Grade 40/20 (Superstructure)	C12G	5350	G/F	R/F	-150	15.940 m²	2.57 m³
R.C. Column (Superstructure)	800 x 600	R.C. Grade 40/20 (Superstructure)	C14G	5350	G/F	R/F	-150	15.940 m²	2.57 m³
R.C. Column (Superstructure)	800 x 600	R.C. Grade 40/20 (Superstructure)	C16G	5350	G/F	R/F	-150	15.940 m²	2.57 m³
R.C. Column (Superstructure)	800 x 600	R.C. Grade 40/20 (Superstructure)	C18G	5350	G/F	R/F	-150	15.940 m²	2.57 m³
R.C. Column (Superstructure)	1150 x 700	R.C. Grade 40/20 (Superstructure)	C2G	5350	G/F	R/F	-150	21.405 m²	4.31 m³
R.C. Column (Superstructure)	1150 x 700	R.C. Grade 40/20 (Superstructure)	C1G	5350	G/F	R/F	-150	21.405 m²	4.31 m³
R.C. Column (Superstructure)	1150 x 600	R.C. Grade 40/20 (Superstructure)	C3G	5350	G/F	R/F	-150	20.105 m²	3.69 m³
R.C. Column (Superstructure)	1150 x 600	R.C. Grade 40/20 (Superstructure)	C4G	5350	G/F	R/F	-150	20.105 m²	3.69 m³
R.C. Column (Superstructure)	800 x 600	R.C. Grade 40/20 (Superstructure)	C20G	5350	G/F	R/F	-150	15.940 m²	2.57 m³
R.C. Column (Superstructure)	500 x 600	R.C. Grade 40/20 (Superstructure)	C38G	3800	G/F	New Flat R	-150	8.520 m²	1.08 m³

Figure 12.8 A portion of the columns (superstructure) schedule.

floor slab. The volume measurements are revealed, corresponding to their geometry sizes. However, special attention should be paid to the area quantities. For concrete walls, only one internal or external face of a wall is counted. If the value is used as a formwork quantity in BQ, other vertical faces must be added where appropriate. In contrast, the area of a structural

Table 12.1 Revit model quantities explained

Object (dimensions)	Area	Volume
Concrete wall (1 m × 0.2 m × 1 m)	1 m²	0.2 m³
Concrete structural column (1 m × 1 m × 1 m)	6 m²	1 m³
Concrete floor (1 m × 1 m × 0.2 m)	1 m²	0.2 m³

column covers six faces including the top and bottom. However, top and bottom formwork normally are not necessary. Thus, understanding of the model quantities is an essential consideration.

The model quantities are also affected by how a model is constructed. Figure 12.9 illustrates four different methods of modelling a simple structure with columns, beams and a slab. In method 1, the four columns rise from the ground floor to the second floor. The four beams are connected to the columns on the first floor and the slab is placed in the void, bound by the beams. The primary problem of this method is that the floor area does not include the beams. In method 2, the slab overlaps the beams but some concrete portions are double-counted. This problem is rectified in method 3, where the beams' top level is reduced to the slab's soffit level. Nevertheless, the model column and slab quantities still cannot be used directly in the BQ. This is because suspended slabs are measured over all construction features, including beams and column heads, according to the Hong Kong Standard Method of Measurement (SMM) for Building Works. On the basis of this, therefore, the four columns should be split into eight columns. Each column starts at the structural floor level and terminates at the structural ceiling level with the slab covering the column heads, as depicted in method 4. This column–beam–slab example explains why the BIM model in this project is so constructed (Figure 12.10).

As many have learned from experience, some model geometries never give the correct BQ quantities irrespective of modelling methods. The mini-piles shown in Figure 12.11, for instance, transfer the loading from pile caps to bed rock. Although the model quantities in Table 12.2 reflect the actual pile lengths, the BQ quantities measure the length of each pile from existing

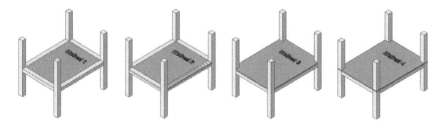

Figure 12.9 Methods of modelling of columns, beams and a slab.

Figure 12.10 A close-up of columns, beams and slabs in the BIM model.

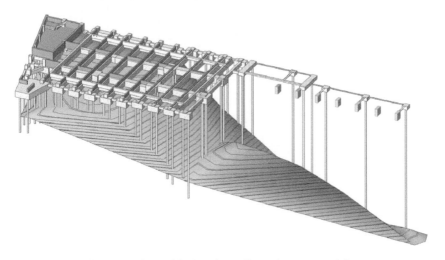

Figure 12.11 The mini-piles and bed rock profile in the BIM model.

ground level rather than pile cap's soffit level. The missing parts (part I and part G in Figure 12.12) in fact do not and should not physically exist and therefore are never taken off from the model. The mini-pile model quantity schedule was exported to a spreadsheet, on which part I and part G were added in order to achieve the required BQ quantities. In short, the default material schedules list only the objects which appear in the model.

According to the SMM, the total length of rock sockets should be given as an 'extra over' item in the BQ (Figure 12.13). Again, no rock socket pile exists in the model, and the quantity is also determined in the spreadsheet.

Table 12.2 A portion of the mini-pile model quantities

2. Mini-Piles

Family	Type	Mark	Base Level	Base Offset	Top Level	Top Offset	Length
Mini-Pile	273mm	24P1	Cap Level 62.23	–5830	Cap Level 62.23	–1200	4.630
Mini-Pile	273mm	23P1	Cap Level 62.23	–5530	Cap Level 62.23	–1200	4.330
Mini-Pile	273mm	25P1	Cap Level 62.23	–6630	Cap Level 62.23	–1200	5.430
Mini-Pile	273mm	26P1	Cap Level 62.23	–7730	Cap Level 62.23	–1200	6.530
Mini-Pile	273mm	26P2	Cap Level 62.23	–8030	Cap Level 62.23	–1200	6.830
Mini-Pile	273mm	27P1	Cap Level 62.23	–8930	Cap Level 62.23	–1200	7.730
Mini-Pile	273mm	22P1	Cap Level 62.23	–7930	Cap Level 62.23	–1200	6.730
Mini-Pile	273mm	22P2	Cap Level 62.23	–7430	Cap Level 62.23	–1200	6.230

:8
Cap Level 62.23: 8

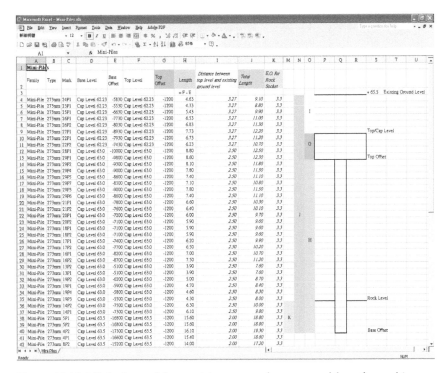

Figure 12.12 Mini-pile model quantities exported to a spreadsheet for making up BQ quantities.

Apart from using the spreadsheet, the same results can also be obtained from within the model quantity schedule. The trick is to create two additional parameters, namely existing ground level and rock socket length, in the master mini-pile. The two constant values, +65.5 metres for ground level and 3.3 metres for socket length, are input and hence automatically apply to instances of the master mini-pile. When creating the mini-pile material take-off schedule, add the rock socket field into the schedule and create a calculated field which can derive the missing length from existing ground level to pile cap soffit level using a formula. Another more automatic method involves the use of application programming interface (API), but the details are not discussed in this chapter.

6 Conclusions

The advent of BIM has led to new thoughts and practices in the construction industry. Yet computerization does give rise to concerns on the future value of some labour-intensive tasks. Material take-off is one of those which can be performed automatically and accurately using BIM. It was observed that the majority of practitioners who were introduced to BIM for the first time had a

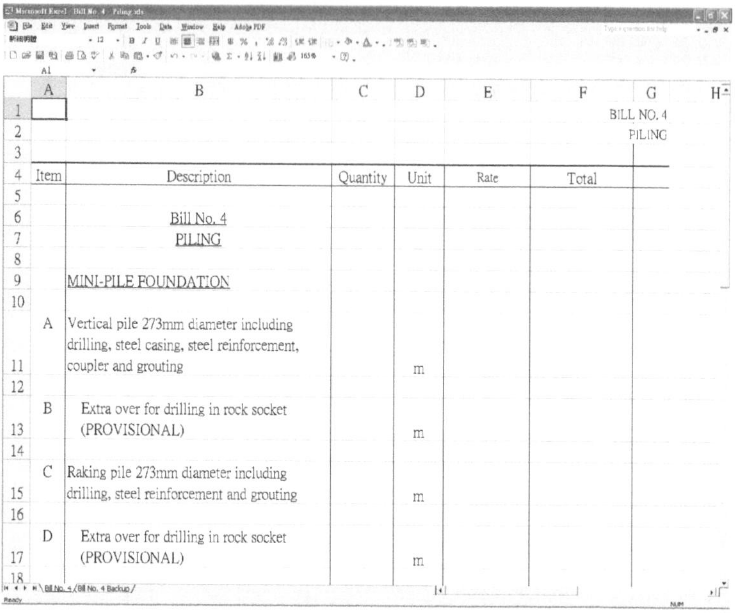

Item	Description	Quantity	Unit	Rate	Total
	BILL NO. 4 PILING				
	Bill No. 4				
	PILING				
	MINI-PILE FOUNDATION				
A	Vertical pile 273mm diameter including drilling, steel casing, steel reinforcement, coupler and grouting		m		
B	Extra over for drilling in rock socket (PROVISIONAL)		m		
C	Raking pile 273mm diameter including drilling, steel reinforcement and grouting		m		
D	Extra over for drilling in rock socket (PROVISIONAL)		m		

Figure 12.13 The 'extra over' items in the piling BQ.

common perception that BIM would take up the role of quantity surveyors. The study in this chapter addresses this perception based on the experience of using BIM in a project in Hong Kong, where the quantity surveying profession and SMM exist. The illustrations and discussions pinpoint the differences between model quantities and bills of quantities. An understanding of how the model quantities are derived by the built-in material take-off function is essential in the first place. The objectives of using BIM should be carefully considered. If material take-off is included, the method of creating the models should take into account the SMM requirements. The default material take-off schedules can list only the model objects which physically exist in the model. A certain degree of manual input is still required to obtain the quantities for these non-existing items. This chapter provides an example of how this could be done by using a spreadsheet or defining additional parameters in the model objects. A more advanced approach could be achieved using the application programming interface (API). In summary, BIM is a tool to facilitate material take-off and not a quantity surveyor. The domain knowledge of quantity surveyors is of vital importance for generating bills of quantities from building information models. However, it may not be the case in some countries where the quantity surveying profession does not exist and the method of measurement is more flexible.

Acknowledgements

The authors would like to thanks Fukien Secondary School and the consultants for providing the necessary information for presentation in this chapter. This chapter also presents part of the findings from research supported by the Research Grants Council of Hong Kong (No. PolyU 5301/06E). This swimming pool project in BIM application has received the 'BIM – A New Chapter of Building Industry' award 2007 from Autodesk.

References

AIAB (ed.) (2006). Autodesk Industry Advisory Board homepage, http://www.aiab.org

Autodesk (ed.) (2007). AutoCAD homepage, http://www.autodesk.com

Bentley (ed.) (2007). Bentley homepage, http://www.bentley.com

Day, M. (2005). 'BIM and the Freedom Tower', *AEC Magazine*, November, http://www.aecmag.com

Gehry Technologies (ed.) (2007). Gehry Technologies homepage, http://www.gehrytechnologies.com

Graphisoft (ed.) (2007). Graphisoft homepage, http://www.graphisoft.com

GSA (ed.) (2006). *GSA Building Information Modeling Guide Series 01: Overview*. Washington, DC: United States General Services Administration.

Hong Kong Institute of Surveyors (HKIS) (ed.) (1979). *Standard Method of Measurement for Building Works*. Hong Kong: HKIS.

Nemetschek (ed.) (2007). Nemetschek homepage, http://www.nemetschek.com

Professional Green Building Council (PGBC) (ed.) (2006). Hong Kong PGBC homepage, http://www.hkpgbc.org

Tse, T. C., Wong, K. D. and Wong, K. W. (2006). 'Building Information Modelling: A Case Study of Building Design and Management', Proceedings of the 31st AUBEA Conference, University of Technology Sydney, 11–14 July 2006.

Tse, T. C., Wong, K. D. and Wong, K. W. (2007). 'Design Visualisation and Documentation Using Building Information Modelling: A Case Study', Proceedings of the 11th CAADRIA Conference, Southeast University, Nanjing, China, 19–22 April 2007.

13 A data-centric, process-orientated model for effective collaborative working

Matthew Bacon

1 Introduction

Proposition: Effective collaborative working in service-based operations needs to bring together four key resources of people, process, technology and data. When people work together systematically using integrated processes, sharing common data, seamlessly exchanged between heterogeneous systems, an efficient and effective service is likely to be the outcome. This is because the people involved in the process will understand how to work together in a mutually supportive manner where their specific roles and responsibilities have been clearly defined. They will be united by a common sense of purpose in a common process. It will be through the common process that data will be able to flow, unhindered by different technologies and platforms, so that all team members receive the data at the point of need, thus enabling them to carry out their part in that process.

In this chapter the author considers how this proposition could be applied to the development of building information modelling (BIM) systems, where data is exchanged as part of a managed process within the collaborating team. The author explains how this proposition has been achieved through his work with a major UK contractor, by deploying his business's Smart Information System (SIS), which will accumulate substantial savings for the contractor's business operations over the next two years. Whilst the case study that will be discussed does not relate directly to the deployment of a BIM system, it nevertheless clearly demonstrates what it is possible to achieve through collaborative working between heterogeneous systems and disparate team members, which is central to the successful deployment of a BIM.

Until relatively recently the collaborative working practices of project teams largely centred on the use of project extranets. The information exchanges using this technology have been largely focused on document-based exchanges, as distinct from data-based exchanges. It is because extranets are unable to manage business processes (indeed, they were never designed to) that they simply function as repositories of information. Unable to manage these processes, project extranets allow team members to maintain their own discrete working practices. Consequently, the kind of collaboration envisioned by the author's proposition would be unattainable with current extranet technology.

Perhaps the users of project extranets would argue that they are obliged to adopt discrete working practices because of the limitations of the software applications deployed in their businesses to function seamlessly with the extranet (Howell and Batcheler n.d.). Whilst there is some validity to this point of view, it ignores the fact that collaborative working, deploying integrated processes, is primarily concerned with the sharing of data throughout the process delivered to the point of need whenever and wherever the data is required.

The emergence of BIM systems means that a new paradigm for collaborative working is possible. Yet proponents of BIM recognise that collaborating teams, in a similar way to that already explained in connection with project extranets, also continue to work in a loosely coupled manner, where they pursue their own discrete working practices and where information is most often shared only at key project milestones (Howell and Batcheler n.d.; Facility Information Council 2007). By working in this way these teams fail to realise the efficiency and commercial benefits that the BIM can bring to collaborative working for both themselves and their clients. This lack of willingness to share information has been linked to information process maturity (Hackos 2006) and is typical of what Hackos refers to as Level 1 maturity (information developers work independently, designing and developing their content in isolation from other developers in their organisation) or Level 2 maturity (information developers occasionally coordinate their efforts to avoid producing the same content more than one time, and occasionally find opportunities to share content developed by other team members, typically through a cut-and-paste process). This is in contrast to the highest level of collaboration, which Hackos refers to as Level 5: where information developers regularly collaborate with colleagues from other parts of the organisation, encouraging a free flow of information and frequent interactions. Whilst Hackos's work is related to document-centric collaboration, the maturity levels could equally be applied to a data-centric collaboration.

It is clear that, without process change and indeed without putting a process at the heart of collaborative working, the adoption of a technology solution, be it an extranet or a BIM, is unlikely to deliver value to the collaborating team or to the client. The implication of the author's proposition for BIM is that the designers of BIM systems must recognise that a process-orientated perspective is essential. Not only should it support the processes of the design and construction teams, but it must also recognise that there are vital business processes and associated data exchanges that need to take place outside of the BIM and yet be capable of being referenced to it. Consequently, the designers of BIM systems must also recognise that the BIM can only ever be a repository for a sub-set of enterprise data. They must recognise that there are numerous heterogeneous systems, managing complex data exchanges that have their focus in business processes other than those focused on design and construction. In doing so the BIM must be able to exchange data with the systems supported by those processes. Inevitably this is where BIM systems and enterprise resource planning (ERP) systems will need to integrate processes and share common data.

An obvious example of this need is where the BIM is increasingly being used to model construction sequencing (Fischer and Kunz 2004), but it is silent on the capacity of the supply chain to meet the demands of what amounts to a hypothetical construction schedule. Clearly an accurate plan must be informed by the capacity of the supply chain to meet the projected demands of the project. Procurement processes are among many processes that take place outside the domain of the BIM and within an ERP system. A data-centric, process-oriented perspective would suggest that the schedule informed by the BIM should be used to inform the procurement systems at the point of initial tender enquiries. Conversely, these systems should in turn inform the BIM to enable alternative construction scenarios to be considered once data concerning supply chain capacity is available, from such enquiries. The author maintains that only a process perspective is capable of informing the designers of BIM systems where data exchanges need to take place between the BIM and these external systems.

Having set the scene for the need for a collaborative process, this chapter will discuss the major issues that need to be addressed in order that the proposition outlined in this introduction might be realised. This will include using a case study to demonstrate how these issues have been addressed using technologies that have been developed in the author's business.

2 Collaborative work in a process-orientated world

2.1 *Industry context*

> BIM is in use today and is flourishing, but it carries many of the problems of the past. These problems are primarily related to stove piping, since many practitioners are only concerned with their phase of the project and do not recognize their role in the overall lifecycle of the facility.
>
> (Facility Information Council 2007)

This paragraph from the National Building Information Model Specification could not summarise any more succinctly the issues that are addressed in this chapter, because it recognises that 'stove piping' or 'silo-based working', as it might be referred to in Europe, where there is little process integration, is a legacy issue that undermines the very essence of collaborative working.

2.2 Process integration

No amount of technology will change the propensity for construction professionals in 2008 to pursue their discrete 'silo-based' working practices. We have seen, in both extranet and BIM solutions, that the most common working practice is one of silo-based working where team members keep resolutely to their own working practices. The document-based paradigm is so entrenched in these practices, and the data-centric paradigm relatively unfamiliar, that this should not be unexpected. In the former example, users provide only the information that they wish to share with their collaborators, when they have to do so. In the latter example, users share data automatically as determined by the data-sharing protocols established by the collaborators. Years of professional practices have become entrenched in what could be referred to as the 'old process paradigm'. Users put up objections in moving to the 'new process paradigm' of integrated working practices, citing reasons such as loss of intellectual property rights and team members being able to access data or information that they are not entitled to. Whilst it may be tempting to suggest that such opinions are based on either prejudice or ignorance, perhaps there are stronger underlying causes. Recent research (Gratton and Erickson 2007) found that the greater the proportion of experts a team had the more likely it was to disintegrate into non-productive conflict or stalemate. The research also concluded that, as teams become more virtual, collaboration declines. And what did they conclude was a significant factor in successful collaborative working? Understanding role clarity was found to be essential. They concluded that collaboration improves when the roles of team members are clearly defined and understood. Now a role-based approach requires a deep understanding of process. In a collaborative process, the roles that team members perform will need to be determined by the tasks that they are required to perform. Consequently a role-based approach and an integrated process are inexorably linked. As we shall see from the case study which follows, the benefits of process integration are substantial.

2.3 Process integration: case study

In the author's own business the contract forming the basis of the case study was to seamlessly integrate a process between a supplier and a main contractor. In this next section it will be explained how this was achieved and thereby enabled the client (the main contractor) to achieve the desired level of process integration between the two businesses.

Whilst the process described in the case study is concerned with plant hire operations, it is relevant to the above discussion on the BIM because it demonstrates how it is possible to dismantle silo-based working practices and replace them with integrated working practices.

2.3.1 Background: haemorrhaging information

The main contractor had established a sole supply agreement with one of the largest plant hire companies in the UK. The supplier was very much disposed to working in an integrated manner with the main contractor and was keen to explore the improvements that this could bring to its own operations.

The established working practice between the two businesses was essentially a paper-based (facsimile and goods receipt notices) information exchange. The problem with it was that documents were continually lost and consequently, when plant hire disputes arose, the paper trail was incomplete. Furthermore the plant supplier recorded most (but not all) information exchanges on its plant management system (to suit its needs), but the main contractor had no such system other than files of facsimiles and handwritten records in a diary. In this one-sided information relationship the contractor often had to pay the supplier for plant hires that supposedly had overrun or for 'lost' or damaged equipment. This situation involved numerous disputes and strained the partnering relationship, because trust was being eroded.

2.3.2 'Joined-up processes' – clear roles and responsibilities

In discussing the issues with both parties a 'joined-up' process was proposed, where data would flow through the process in such a way that each person involved in the process received the information that they required relevant to their role in it. (This proposal was very much in line with the proposition stated at the outset of this chapter.) The strategy was that there would be no duplication of data – only what we refer to as the 'one instance of the truth', where data is stored once but reused many times. Furthermore, because the existing process was incomplete, in that it addressed only specific process phases, a complete plant hire process addressing all phases of the plant hire operation was developed, where all the participants in it have clearly defined roles and responsibilities in the process (Figure 13.1).

The new roles that emerged were roles such as 'plant administration', 'purchase authority' and 'site buyer'. These roles were performed by people with existing job titles, called 'plant centre manager', 'project manager' and 'foreman'. In this 'new process paradigm' anyone with appropriate training could perform one or more of these roles. Indeed a person could have one role on one contract and another role on another contract. The system enables them to interact with the process based on the role that has been assigned to them.

2.3.3 Making process knowledge explicit

The process that emerged considered the role of each person to be involved in the process and how data would flow from one system to another as each

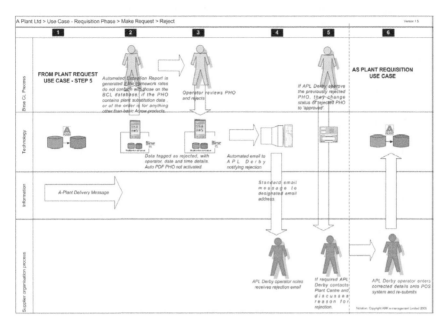

Figure 13.1 Modelling collaborative roles in a process.

Notation developed by the author.

part of the process was enacted. Key to enabling data to flow was an XML (open standard) message schema that would transport data from one system to another during the process. In this case the supplier would initiate proposed orders on its system, having received a request from a construction site and initiated a proposed order; the messaging system would be activated and send the request to the contractor's system (the author's business's Smart Information System), where it would then be subjected to a validation process, supported by a 'rules engine'. (This is the contract validation between the two collaborating parties referred to earlier.) This rules engine parsed the data content in much the same way that a plant manager in the 'old process paradigm' would review fax information content generated from the plant supplier's system. This was achieved by documenting these rules with the plant manager and configuring the 'rules engine' accordingly. If the validation process approved the message, the next stage of the process would be initiated, and a new message would be created with a plant hire order number inserted that would be returned to the supplier. At this stage a draft plant hire order would be visible on the system. However, if the original message was not approved then the process engine would halt the process and issue a 'process event' warning to the plant administrator role. At this stage the system would advise the user exactly why the process was halted. The user could then speak to the supplier and discuss the issue, and either reject the proposed data flow or accept it. If the user accepted, the process engine would then automate the next stage of the process.

2.3.4 Learning outcomes

- *The benefits of clear role definition.* The role-based approach was fundamental to the success of this system. It enabled the main contractor to spread the workload across more people in the team. This resulted in a greater capacity to carry out work with the same resources. Today the plant centre has tripled its workload, but the number of people managing the work is still the same. The construction sites have fully embraced the technology, because it makes life much easier for them. Notably they get plant on-site more quickly because the processing of plant requests is quicker and, as we shall see in the next section, they also get 'rich', accurate data about all plant hire status automatically delivered to them each week – what the author refers to as 'information delivered on a plate'.

  The role-based approach has also led to process efficiency, because team members can be assigned roles with different levels of authority. This may well be a requirement for larger, more complex projects, but perhaps not needed for less complex projects. The process requirements will determine who has authority to approve certain types or certain status of information. In the context of a BIM this means that the role of a design manager, for example, can be clarified into role sub-sets, so that a design brief manager (who controls the BIM project brief) may be responsible for managing the automated brief compliance technology and will also have specific access rights to information and specific rights for approving it, whereas a design coordination manager, who is responsible for interface management, may be responsible for managing the automated BIM code checker and following through non-compliances with the team. By breaking out roles in this way, the workload can become more easily distributed, with clear lines of responsibility.

- *The benefits of integrated processes.* Without further explaining the case study process it is hoped that it can be appreciated how an integrated process between two parties can be implemented and most importantly how, through an open standard transport system, data can be passed between heterogeneous systems. Importantly it was the process that defined the required data flows, and in the author's opinion it is this premise that needs to be applied to the design of a BIM system. In this regard there are two process perspectives, the processes that need to take place between the participants working within the domain of the BIM system, and those participants who are peripheral to the processes on which the BIM is founded but who will either make use of BIM data or add to it.

In each instance the benefits that have emerged are that the whole process was speeded up so that not only was plant delivered to site more quickly but there was a clear, auditable data trail between the supplier and main contractor that was previously missing. A significant benefit to the main contractor of this visibility of data was that it became immediately obvious where the supplier had been overcharging the main contractor and had been adding expensive accessories to plant hires which had been previously unaccounted for.

In terms of the benefits which emerged to participants in peripheral business processes, the system is able to initiate data flows to inform people involved in other processes, such as financial management and supply chain management, of key process events. Presently these people are automatically sent emails when these events arise, where either a PDF or a spreadsheet file is attached containing the information that they require. The need for data flow is the subject of the next section of this chapter.

3 Collaborative working in a 'data-centric world'

3.1 *Industry context*

In order for a BIM to be fully implemented and its potential fully realised, it must allow for the flow of information from one phase to the next. This can only be achieved through open standards.

(Facility Information Council 2007)

Again this paragraph from NBIMS clearly articulates another issue central to the argument set out in this chapter:

- Achieving information flow is essential within a process comprising sequential phases.
- Open standards for data flow are essential.

It is because the BIM has the potential to be able to exchange granular-level data as a transactional process between heterogeneous systems, rather than as aggregated data exchanged as a batch process, that achieving information flow in concert with an integrated process is distinctly possible.

3.2 *Data flow*

Bentley and Workman (2003) discuss the basic needs of BIM, and identify the need for data management as a key requisite. Fundamentally they call for synchronised, shared, context-sensitive information delivered to the point of need for informed decision making. This is what the author would define as a data flow, where data (information) flows to the point of need. In this context it is also relevant to note that the authors of NBIMS state that such flows (data exchanges) must take place using open standards.

In the collaborative world of integrated processes, achieving information (or data) flow is essential if the effort expended in defining such processes and the costs of investing in the technology are to be recouped. The return on investment (ROI), as it is referred to, must demonstrate greater efficiency and a high level of predictability in the delivery and quality of the output. If participants are waiting for data or information from upstream activities and find, when they receive it, as Bentley and Workman describe, that it has not been synchronised, that is, coordinated with other data in the BIM, then flow in the process will be interrupted and delays will ensue, thus eroding the ROI. Consequently, reliance on an open standard approach alone, as advocated by the authors of NBIMS, to facilitate data flow in the process is not enough. Data flow must be managed and it must also be synchronised.

That the adoption of open standards is of central importance is unlikely to be disputed. However, there is much room for debate as to how such standards should be used. Most BIM system designers would no doubt advocate that they should be used in the data transport mechanism from one system to another. However, in the author's experience there is also much to be gained from using open standard XML databases, which are inherently more flexible than relational databases. Indeed the benefits are substantial when they are used together as part of one system. This is because XML databases are very good for storing meta-data referenced to data in the relational database. XML provides the contextual information so necessary when considering data flow through a whole lifecycle process.

Bentley and Workman (2003) assert that 'context' is also of fundamental importance when managing data flow. Users of the data will need to know in what context the data was created, and this will determine how it could then be used upstream in a later process phase. Context also provides data with meaning. In the case study we will see that meta-data is applied to plant hire data in different ways. For example, plant asset data, such as asset codes, asset descriptions and physical properties, is stored in a relational database. But data on which site ordered the plant, who ordered it, the contract number that it is being charged to and whether it was damaged when it arrived would all be stored in the XML database. Meaningful information about a specific asset can then be compiled from the relational database which supports the asset management system and the XML database that stores the contextual data.

In the case of a BIM which is managing data across a facility lifecycle, being able to distinguish 'as built data' from 'design data' in a facility management context would be paramount. Consequently management of data flow is also concerned with being able to ensure that *relevant* data flows from one process phase to the next in order that it can be reused. All too often it is being re-created by downstream teams, and this is a terrible waste of resources and significantly erodes the potential ROI.

3.3 Data flow: case study

One of the key objectives of the project forming the case study in this chapter was to manage data flow to the point of need. Procurement managers and regional buyers, quite disconnected from the plant management process, needed to understand issues such as the volume of work being passed through the supply chain. Their need was to drive out transaction costs and yet ensure that sites received the best possible service from the supply chain.

3.3.1 Silos of information

A silo-based perspective, which characterised working practices prior to the deployment of the SIS, meant that all of the purchasing information (such as it was) was held by the plant management team. The procurement team (in another 'silo') had no access to this information, and consequently poorly informed purchasing decisions were being made through a lack of information being available where and when it was needed.

In contrast, regional buyers also maintained there own 'silos' of data in spreadsheets. Each region maintained its own data about suppliers, which was not shared with the plant management team. Consequently, the plant management team had to develop their supplier information, and this was clearly of no value to the business as a whole. In considering other processes, such as estimating and tendering and financial processes, again each 'silo' had supplier information to support its processes. Clearly there would be benefits in sharing supplier data across large parts of the business.

3.3.2 'Joined-up processes' – 'joined-up data'

The project addressed this challenge by creating a comprehensive supplier database, and storing all purchasing data associated as meta-data for each supplier. Because the data is stored at a highly granular level, it is possible to determine not only what volumes of transactions are being entered into, but what volumes are being managed for specific categories of purchasing. Furthermore these categories are coordinated with the financial system, so that financial breakdown structure and plant/material breakdown structures are compatible. It is important to ensure that users with roles not directly associated with the core process being managed by the system are also able to access the data for those processes that surround it. These data flows extend to the suppliers in the supply chain from which plant and materials are procured. Part of the system has been opened up to them so that they are able to maintain their supplier records directly on the main contractor's system. This has saved the contractor substantial administration overheads, as well as ensuring best possible accuracy of supplier data.

3.3.3 Achieving flow

The system that was deployed between the main contractor and the plant supplier was dependent on a resilient data flow between both parties. For information to flow, then data has to flow. The data has to flow in concert with the process. As each business works through a process and each person performs a role in that process, the system is exercising control over the process.

From the supplier's perspective, only users with specified roles could generate a plant hire order request, and the supplier's system was required to transmit certain specific details of the person performing the data entry with the message that flowed to the main contractor. The supplier was also required to use specified equipment codes and accessory codes. These codes related to the equipment that the supplier was authorised to supply to the main contractor, and the accessory codes controlled the quality (and therefore cost) of the accessories to be supplied with certain pieces of equipment. As was explained in section 2.3.3, when the message containing this data was received by the main contractor the message content was subjected to a process where it was effectively scrutinised (what I referred to earlier as 'parsing'). If data content (contained as an XML file) was found to be incorrect then the message was rejected and the supplier notified. The notification contained detailed information as to why the transaction had been rejected. However, if it was approved then, as stated earlier, it was passed through the next stage of the process.

3.3.4 Data synchronisation

On many occasions the main contractor was not aware of the orders passing through the system (and had no need to know when each transaction was being processed). In these situations the first notification that it would receive from the system was if a delivery had not been recorded on the system within a specified period of time from the accepted place order (48 hours). Should that happen then an event management system would notify it of the failure and it could, with a few mouse clicks, ascertain which supplier was at fault and take the appropriate action. This is what the author refers to as a 'pull system', where users are in control of the data that flows to them.

One way in which the data was synchronised on the project was to configure the plant supplier's system, so that every change that was made to the hire contract initiated a message that was then automatically processed by the main contractor's system. For example, the supplier was notoriously poor in off-hiring plant despite repeated requests from the plant centre to do so. The longer the equipment stayed on-hire the more money the supplier made. With the integrated processes, the system registered a 'suspension request', and if an off-hire number was not received on the system within

24 hours this raised an 'event' on the system warning the plant management team that the supplier was in default. All these transactions are displayed in a management interface called 'Plant Tracker' (Figure 13.2).

However it was not only the plant administrator who required information, but the construction sites too. Information for them is automatically compiled and delivered by the system to each site every Thursday night in the form of a spreadsheet attached to an email. This document contains all the latest status data for the plant on their site. It also highlights plant hires which are overdue for delivery or collection (cells coloured red) as well as those that are nearly overdue (cells coloured amber). This is what the author refers to as a 'push system', where the system behaves as configured and automatically delivers the information to users with specified roles according to their needs.

The system also monitors the service level agreements (SLA) for each supplier and also automatically issues an automated weekly performance report, with data processed from the database. Another example of an automated 'push system' report is where any major transactions on the system are automatically logged by it. For example, if a user changes supplier details or maybe the contract details then in this instance the system automatically creates an audit report for the system administrator. Should there be any disputes, all such events can be readily traced to the person performing the work, including the exact date and time of the event (metadata). As will be explained in the next section, the automated reporting functionality in the Smart Information System has saved substantial administration time for the plant centre, and these savings have contributed

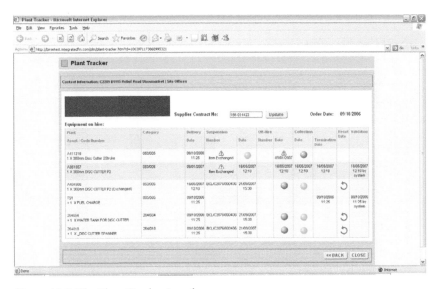

Figure 13.2 The Plant Tracker interface.

to the ROI in terms of being able to manage larger volumes of work without the need to employ further resources (see Figure 13.3).

3.4 Learning outcomes

• *The benefits of automation.* Automation of repeatable business processes has demonstrated substantial efficiency savings. During a two-year period the plant hire process alone has achieved cost savings amounting to a little over £0.75 million. It has also enabled users to develop significant trust in the system. This is because the information exchanges have consistently proven to be correct, because there is a clear and unambiguous data trail to support them. The system has encouraged much wider collaboration between disparate team members (in different silos). This is because all of them now receive the information (or data) that they require to perform their job roles, so that they can focus much more of their attention on what this means to their collective work rather than having to spend time attempting to find the information or indeed questioning its authenticity.

Automation delivers 'data flow' into the process, and this means that users get the information that they need, when they need it, in the form in which they need it. This makes decision making much more efficient, with greater predictability in delivery of service. However, automation must remain in the control of the users. Clearly it is untenable for a process to be able to run without users being able to intervene. This was found to be a very important feature of the system. Consequently, if a data exchange is challenged, it can be halted, investigated and if

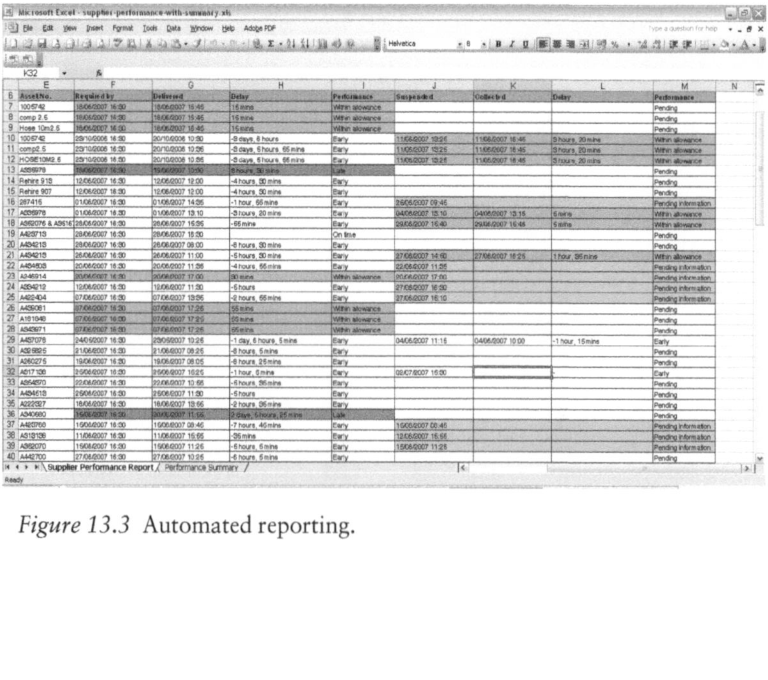

Figure 13.3 Automated reporting.

required manually re-run. Only certain users should have the ability to manage the automated process in this way, because without such control it would be easy for the system to be abused.

- *The benefit in having visibility of process status.* Having visibility of data flows also provides visibility of process status. For example, the team specified that once data achieves a certain state then a process control gateway could be triggered (to enable a decision-making process, for example). If the team understands what information is required to make a decision and on what data that information is to be based, then it follows that visibility of the state of the data is an essential requirement of a system.

In the case study system, process status determines:

- when orders can be placed on contracts;
- when SLA performance measurement is triggered;
- what point in the on-line supplier registration process a supplier has reached, thus controlling when orders can be placed with it;
- at what point in the process purchase order authorisations are required.

The author stated earlier that in his opinion the BIM offers a 'new process paradigm'. The foregoing examples illustrate very clearly why this is the case. It is because BIM technology is now largely database-driven technology (as distinct from the 'closed file' structures of previous technologies) that data flows can now take place at a transactional level of abstraction rather than at a batch level. This is analogous to just-in-time data delivery and lean processing. For collaborating teams using BIM technology this could mean, for example, that the quantity surveyor could be notified when quantities or specifications change in the BIM after a certain process status has been reached. It could also mean that the design coordination manager is notified when fire compartment volumes change in the model, so that action can be taken to assess the effect of such proposed changes.

- *The benefit of open standards.* Facility Information Council (2007) states that open standards for data flow are essential, and the author would clearly endorse this statement. The case study proves that information flow between heterogeneous systems is possible by using such standards, which in this case was an open standard messaging system built using Java and XML. This is a significant business benefit.

The transmitted XML data was then stored in an XML database and, as its state changed through the process, mark-up was added to the data in the database, enabling a history of transaction changes to be stored with it. Each time the data needed to be compiled into reports, the system (an information processing engine) would transform the data and assemble it into an html, PDF or Excel output appropriate to the needs of the user. The ability to extend the properties of data sets in the

XML database, and thus provide 'rich content' to the business, is also of great benefit.

4 BIM and collaborative futures

There is not one enterprise technology solution that could be deemed 'state of the art' which does not have a process engine technology driving it. Process engines manage and automate repeatable processes, freeing up the organisation's workers to focus on value-added activities and eradicating waste. In the author's opinion BIM technology should be considered part of an enterprise suite of products. Designers of BIM systems should recognise that a process-enabled BIM is fundamental to achieving effective and efficient working practices not just within the immediate domain of the BIM but for all those other enterprise processes that need to function around it.

Enterprise middleware infrastructure now acts as the technology glue that binds heterogeneous technologies into an integrated system. The Oracle Fusion® and the SAP Netweaver® middleware platforms are examples of these technologies. At their heart is an 'enterprise service bus' (ESB) that provides the messaging system between these technologies and binds together services to support the enterprise, enabling data to flow and processes to be integrated.

In analogous terms we can think of the ESB as the motorway network for the enterprise and each of the enterprise systems, such as financial, procurement and personnel, as well as the BIM, as being the industrial parks where work takes place and which are connected to that motorway network by interchanges such as slip roads and roundabouts. Each enterprise system, therefore, requires a connector to the ESB, and this is where BIM system designers need to recognise that, in order to achieve a process-orientated, data-centric solution that binds people, process, technology and data into a 'system', they also need to develop connectors to the ESB (Oracle currently use an Application Integration Architecture (AIA) to achieve this).

BIM system designers also have to recognise that, in a collaborative future, no significant technology can exist in isolation, because it will be overtaken by technologies that do support collaborative working across an extended enterprise and not just the rarefied processes of a design or construction team. Today, the majority of the largest contractors have invested in either Oracle or SAP business systems, and they will be looking to understand how BIM technologies can be leveraged on the back of them. They are seeking to understand how they can break down the 'silos' of discrete working practices and adopt new working practices that eradicate waste, improve productivity and add value to the organisation. Indeed, as the case study has demonstrated, they are indeed actively investing in technologies that are both process orientated and data-centric.

References

Bentley, K. and Workman, B. (2003). 'Bentley Systems Inc.: Does the Building Industry Need to Start Over?', http://www.laiserin.com/features/bim/bentley_bim_whitepaper.pdf

Facility Information Council (2007). *National Building Information Modelling Standard, Version 1.0: Part 1 – Overview, Principles, and Methodologies*, December. Washington, DC: Facility Information Council, http://www.facilityinformationcouncil.org/bim/ (accessed January 2008).

Fischer, M. and Kunz, J. (2004). 'The Scope and Role of Information Technology on Construction', CIFE Technical Report No. 156, February, Stanford University, CA.

Gratton, L. and Erickson, T. J. (2007). 'Eight Ways to Build Collaborative Teams', *Harvard Business Review*, November, http://harvardbusinessonline. hbsp. harvard.edu/b02/en/search/searchResults.jhtml;jsessionid=YETP0JJBJCRKCAK RGWDR5VQBKE0YIISW?Ntx=mode+matchallpartial&Ntk=Author%20Name &N=0&Ntt=Lynda+Gratton

Hackos, J. T. (2006). *Information Development: Managing Your Documentation Projects, Portfolio, and People*, p. 31. Indianapolis, IN: Wiley.

Howell, I. and Batcheler, B. (n.d.). 'BIM: Building Information Modelling Two Years Later', http://www.bim.utah.edu/Website%20Info/Articles/newforma_bim.pdf (accessed January 2008).

14 Supporting collective knowledge creation in digitally mediated collaborative environments

Tuba Kocaturk

Investigating a new and evolutionary design domain is a challenging task which requires the selection of a critical strategy with an awareness of the possible contradictions between past understandings and emerging characteristics about design. There is a critical balance between the extent to which we allow established preconceptions to influence our inquiry and the extent to which we are open to the emergent concepts that will challenge established understandings about design and design knowledge. This has been an initial motivation for our research, which has led us to the following question: 'How can we conceptualize the digitally mediated design domain in order to understand and identify the knowledge content it entails specific to its unique context?' Any attempt to answer this question necessitates a recognition of the key themes which distinguish the new domain from the conventional designs:

- the extent to which digital technologies are integrated into the design and production processes;
- the emerging formal/tectonic qualities and varieties;
- the changing socio-organizational roles and responsibilities of stakeholders.

While each theme introduces new concepts, working processes and skills into architectural design, the definitive lines between the working processes of disciplines start to blur. Moreover, various types of interactions across these themes start to define new dependency types between design tasks within and across disciplines, contributing to the evolution of the final artefact. The multifaceted interactions between the material and mental creation of artefacts and the unprecedented ability to construct various complex geometries and representations (facilitated by the digital media) are the key to understanding the emergent knowledge needs of digitally mediated design. One of the most crucial characteristics of this new field of design knowledge is that it is constructed collaboratively by the various parties taking part in the design and implementation processes. The emerging interactions between the design and production processes become highly non-linear and dynamic, leading to the emergence of a new, cross-disciplinary and collective body of design knowledge.

In this highly digitally mediated context, the emerging design methodologies and the emerging knowledge needs of multidisciplinary teams start to diverge from the conventional models. Understanding the characteristics and the key aspects of digitally mediated design processes and collective creativity is the key to the future developments of the next-generation technologies and to providing the necessary support for collaborative design. This can be achieved not only by facilitating semantic interoperability among disciplines from conception through to production of designs, but also by acknowledging and allowing the coexistence of multiple methodologies and the changing knowledge needs in the building industry.

In the following sections of this chapter, we will attempt to conceptualize digitally mediated design, with a strong focus on the free-form design context. Based on a comparative analysis of cases, the development of a taxonomy and a 'collaborative knowledge framework' will be described (Kocaturk 2006). A knowledge framework is composed of a collection of concepts, principles and experientially verified relationships useful for explaining the design processes. The knowledge framework intends to serve as a reference model to frame and evaluate different design experiences and their associated knowledge. The rationale behind the development of such frameworks is to support the designers in their knowledge acquisition, knowledge construction and sharing with a common structure and shared set of descriptive terms.

Following this, a web-based system has been developed by integrating our knowledge framework into an existing database and by adding supplementary functionalities to its representational structure for efficient access to the related knowledge content. A long-term goal and motivation for the development of this prototype has been to support the collective creation and transfer of design knowledge where new knowledge can be added and retrieved by different design participants. In this system, the growth of the knowledge content is intrinsically dependent on user participation. Using the characteristics of the domain content identified, the prototype aims to provide a flexible structure for the organization and representation of the situated and collaborative knowledge elements.

Finally, the prototype will be evaluated according to the factors that influence its effectiveness, applicability and further development in varying collaborative contexts. Such an evaluation becomes crucial given the fact that the system is intended to grow with user participation and their reflection on design processes, contributing to the collective and collaborative construction of knowledge.

1 Design practice: a comparative analysis of three cases

In current digital design practice, several differences have been observed in the ways and extent to which designers incorporate new tools and technologies into their working processes. Similarly, the ways in which the design and implementation processes are structured and conducted vary from one practice to the next. This chapter reports on the process of developing a 'knowledge framework' based on a comparative analysis of three design projects. Rather than starting with a theory to be investigated, we will report on the process of inducing a theory based on the comparative analysis of cases which represent the diversity and complexity of the knowledge content under study. The analysis of the cases have been guided by a contextual framework comprising the following categories (interdisciplinary phases in a project lifecycle):

- *design intention* – specification of the formal qualities and design approach;
- *representation* – representation and description of the design object for subsequent phases;
- *rationalization* – division of the free-form surface into rational cladding components combined with a supporting structure;
- *fabrication* – selection of the fabrication processes, tools, techniques and strategies;
- *materialization* – decisions concerning the formal and behavioural properties of the structural system, the elements and the choice of materials.

The first case is the Web-of-North-Holland (WNH), a temporary exhibition building built in Harlemermeer, in Holland, designed by the Dutch architectural firm ONL Architects. The second case is the Dynaform Pavilion, a temporary exhibition building built in Frankfurt, designed by the German architectural firm Franken Architects. The third case is the Experience Music Project (EMP), a museum building in Seattle, designed by the American architectural firm Gehry Partners.

2.1 Data analysis, coding and theory building

This section will give an account of how the data collected throughout the case studies have been analysed and conceptualized. The method used for the acquisition and construction of concepts is the manual coding of documents and transcripts gathered on three cases. A content analysis of transcripts for each case aims to identify, explain and compare the unique ways in which different practitioners frame design problems and solutions, the terminologies they use to refer to specific concepts, and the tasks and procedures they follow in unique situations. The coding has been conducted parallel to a horizontal and vertical analysis of cases. With horizontal analysis, the concepts and their interactions with other concepts have been identified within the specific context of each case. In addition to this, the vertical analysis facilitated a comparative study of these concepts across cases – in different contexts of use – in order to understand the hierarchical relationships between concepts at different levels of generality.

Figures 14.1, 14.2 and 14.3 depict a partial transcript analysis of the three cases. Each text depicts how the concepts were interconnected either to formulate or to solve a problem during the design process. In each text, the words and phrases are either clustered into one concept (terms in parenthesis) or, alternatively, are taken as they appear within the text (highlighted text) in accordance with their meaning within the specificity of each context. What each concept refers to (e.g. method, process, task) and its frequency of occurrence in other cases are the two determining factors for the labelling (coding) – the level of generalization used to identify the concept. For example, 'the aligning of the surface and structure' is a concept which is a decision concerning the *rationalization* of the structure and the skin. The comparative levelling of the concepts will be described in section 2.1.3.

2.1.1 Meta-analysis

In each of the three cases analysed, the overall design process displays a continuous negotiation and reconciliation between multiple perspectives utilizing multiple terminologies expressing the needs of a multidisciplinary team. We created a meta-analytic schedule, which is a cross-case summary table in which the rows are case studies and the columns are the variables

The final form is a **continuous double curved skin (FORMAL INTENTIONS)** , which becomes **a frozen instance of acontinuous deformation process(DIGITAL MODELLING)**. The **supporting structure (STRUCTURAL SYSTEM)** was required to follow the exact geometry of the architectural form........... The structural frames were generated **bytaking 16 cross sections from the master geometry(SECTIONAL CONTOURING)** each section taken at a different angle, and each resulting in a unique shape, and each following the outline of the master geometry........... The special **material properties** of the surface material (membrane) and its connection with the supporting structural frames were the main determinants of the **surfacerationalization (CLADDING ORGANIZATION)**........... The engineers translated the structural form of each membrane segment into separate **cylindrical, conical and flat surfaces (RULED SURFACES)**........... A full-size **mock-up (PHYSICAL MOCK-UP)** of several structural frames had been produced. enabling the design and construction team to identify and resolve possible problems with the **connections between the components**. The flanges were all rectangular cut **(CROSS-SECTION)** and had to be **rolled (COLD-FORMING)** according to the curvature of the top and bottom outlines of the boxes and the individual pieces were cut using **computer-driven-plasma-cutters(CNC-CUTTING)**

Figure 14.1 Extraction of domain concepts during case analyses: Dynaform.

The firm has a specific approach to the design and production of free-form surfaces generally referred as "**paper surface forms**".......... The shapes and materials bent into these shapes   **may be formed without the need of stretch forming (DEVELOPABILITY)** .... The surfaces created and constructed in this fashion define both the exterior and interior surface qualities of Gehry's   **double-skin** buildings **(MULTI-LAYER SKIN)** .... The surfaces of the initial models are captured in CATIA by using **3D scanning**, at an early stage.......... The original surface of EMP was rationalized into conformance with the **Gaussian curvature analysis**.......... The fully curved free-form sha   pes of the original design were supported as a composition of Gaussian curvature constrained paper surface forms...........The generative studies **torationalize the surface(SURFACE RATIONALIZATION)** attempted to optimize the layout of the face sheets on individual panels, on the basis of available **material sizes**. Skin options had to be examined not just from a fabrication point of view but also for loading, attachment to the **structure (STRUCTURAL SYSTEM)**, affect on system performance, etc........... **CNC guided plasma cutters (CNC-CUTTING)** were used to cut the flanges of the curving structural steel members. **Shotcrete** and waterproofing were applied to reinforcing steel overlaid with stainless-steel hardware cloth.

Figure 14.2 Extraction of domain concepts during case analyses: Experience Music Project (EMP).

The main decision at the macro level was to create a shape in which the **surface and the structure would be aligned (SKIN/STRUCTURE INTEGRATED)** without the need for a secondary structure............. The final form was composed of a **single skin double curved geometry** which had further been smoothened in respond to **materialization** and performance criteria.......... For the rationalization of the surface, ONL applied a unique **tessellating** system, which is based on **triangular grid (TRIANGULATION)** of an icosahedron (a 20-faced polyhedron).......... A 3D structural grid was created by mapping the icosahedron on the **NURBS** surface..........The choice of this system had further been justified b ecause of the **regularity of connection** details as well as **directional uniformity** it provided..........When **two non-parallel lines which are not in the same plane were combined (DEVELOPABILITY)**, a challenge was presented. At this point rather than combining the lines with a curved surface, a **folding plate** solution was preferred....Every triangular **hylite** element (surface cladding) was also **unfolded** and **water-jet cut**.... Since the corners of the **claddings were bent (COLD FORMING)** manually to fit the geometry, unexpected distortions and displacements occurred at the middle part of these surfaces which caused unexpected changes in the geometry of the overall form.

Figure 14.3 Extraction of domain concepts during case analyses: Web-of-North-Holland (WNH).

Table 14.1 A summary of the case studies with the extracted concepts

	Rationalization	Materialization	Representation	Fabrication	Design intention
EMP	Gaussian analysis. Surface subdivision. Framing.	Sheet metal. Geometry of skin. Geometry of structure.	Surface model. 3D scanning. Point data. Paper surfaces. Master model.	Shotcrete. Plasma cutter. CNC cutting. Forming process.	Paper surface. Multi-layer skin. Physical models.
WNH	Triangulation. Skin structure integrated. Tesselating. Sectional contouring.	Planar elements. Frame action. Primary structure.	Developable. NURBS. Unfolding. 3D scanning. File Exchange Protocol.	Unfolding. Water-jet cutting – cold forming.	Directional uniformity. Single-layer skin. Parametric modelling.
Dynaform	Skin-structure integrated. Irregular subdivision. Curvilinear members.	Steel thickness. Vierendeel truss. Semi-structural membrane.	Ruled surfaces. Wireframe model. Full-size mock-up.	CNC cutting. Manual forming. Machine code. Fabrication tolerances.	Fixed form. Digital modelling. Parametric design. Single curved.

identified for each category. The purpose of meta-analysis is to allow us to use the summary of case studies to make theoretical generalizations. Table 14.1 shows the collected concepts (variables) from each case, under each category, after a comparative study.[1]

2.1.2 Horizontal analysis

The horizontal analysis serves to formalize and compare the strategic and situational interrelatedness of the concepts within and across categories for each of the three projects studied. It is conducted to analyse the interactions of context-specific design concepts and the emerging links between them, which are identified as the main sources of emerging design knowledge. This will facilitate a deeper understanding of how concepts are linked at different levels of abstraction, how strong these links are and how the value, strength and direction of these links vary. The horizontal study is conducted parallel to the vertical study to improve the terminology and sub-category descriptions under each category. While the content analysis aims to extract the concept vocabulary of the domain in each case, the horizontal analysis aims to understand the value, strength and direction of the links between the concepts in each project. The horizontal analysis is directly taken from the content of the text and is graphically illustrated for each case separately, as shown in Figure 14.4.

Following the commonly accepted notion that there is never a complete representation of the design problem in the head of the designers (Lawson

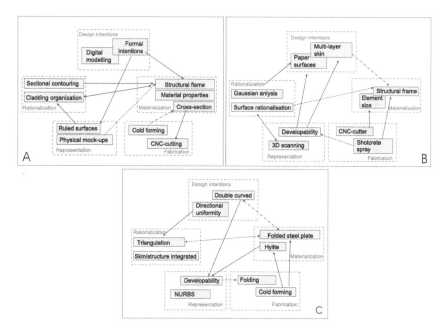

Figure 14.4 Emerging concepts and their interactions (knowledge elements) across various design stages in (A) Dynaform, (B) Experience Music Project and (C) Web-of-North Holland.

1980), in the horizontal study we focused more on the local and non-sequential network of links in different design situations in each case. As has been observed from the case studies, each category comprises a sequence of concepts (strategies, solutions, techniques and methods) specific to each project and their associations. While the clustering of these concepts under the five main categories helps in understanding the purpose of each concept within its context, showing the associations (or links) between them helps in explicating the unique ways in which different practitioners frame the design problems and the solutions they bring to unique situations. The defining characteristic of these design problems is that they are highly interdependent. According to Eastman *et al.* (2001), a better understanding of the process of structuring problems allows further insight into our understanding of the design processes followed by designers.

As shown in Figure 14.4, the emerging concepts under each category, such as 'Cold forming' (A), 'Gaussian analysis' (B) or 'Triangulation' (C), provide the declarative and procedural elements of the emerging knowledge which is referred to as *process* and/or *task* knowledge. It has been observed that new knowledge emerges not only by introduction of new concepts (tools, tasks, processes, etc.) but also by the introduction of new links between concepts which vary in their value, strength and direction. The links created between concepts define the unique ways in which different practitioners frame the design problems. Thus the links and associations created between the concepts can be recognized as the strategic and situational dimensions of design knowledge. While situational knowledge is necessary in recognizing problems for the selection of the relevant concepts, strategic knowledge refers to the conscious and tactful decisions made (de Jong and Ferguson-Hessler 1986). The links may change in both meaning and form with regard to how they are linked according to the viewpoint of the project participant who creates the association between two or more concepts. For example, in EMP, the division of the surface into individual cladding components (cladding organization) has been determined by the available sheet-metal sizes and the constraints coming from the fabricator. This way, a link is created between *rationalization* and *materialization*, which reflects the approach of the design team and affects the choices made and the evolution of the final artefact. Alternatively, the organization of the cladding components could have been determined solely by the aesthetic criteria. In a similar example, the membrane surface in the Dynaform project is represented by *ruled surfaces* (under *representation*) for ease of constructability. The use of a variety of ruled surfaces is further justified to create an impression of a double-curved overall surface to comply with the formal intentions (under *design intentions*) of the architect.

Different links have been created between the same two concepts by the different members of each design team. For example, while the cladding organization in EMP is conducted by the architectural team, the same process becomes the task of the engineering team in Dynaform, which affects the ways in which problems are formulated and solutions are generated. In this respect, understanding the viewpoints of the people in the collaboration – whose personal focus based on their disciplinary background might influence their decision process and choices – is essential for understanding the ways in which the form has evolved into its final shape. While certain relationships are emphasized more in one project, others become less important or are totally ignored as the design process progresses. With each project, new relationships are introduced, either empirically or through the introduction of new facts and relations. Furthermore, the links change in both meaning and form, reflecting the viewpoint of the project participant who created the link. The change in meaning relates to the dependency relation types (e.g. constraint, influence, inspiration) between different aspects, whereas the change in form relates to the change in the direction of dependencies (e.g. bi-directional, mono-directional). An additional observation concerning the dependencies between concepts is that they can be linked at any phase of the design process even though they belong to different phases of design. For example, constructability criteria can influence the form generation process even at the very early stages of the design. Similarly, a specific production technology (e.g. CNC-cutting) will influence the maximum thickness and size of the cladding materials that can be processed by using this specific technology and will in turn inform the organization of the cladding components.

The case studies have also contributed to an understanding of the radical differences between collaborative and individual creation of knowledge. The highly non-linear process of knowledge exchange between stakeholders facilitated the creation of shared meanings between the members of the design team. The resulting knowledge is defined as 'collaborative', which is constructed through the interaction of multiple actors and embodies the dynamic elements of knowledge that would be difficult to generate by an individual. This view propagates a rather process-oriented view, and is key to the understanding of the knowledge content of the digital design paradigm in architecture.

2.1.3 Vertical analysis

Vertical analysis aims to cluster the concepts into sub-categories in a hierarchical organization. The concepts extracted and collected during content analysis for each case are compared with the other concepts across cases. Based on this comparative analysis, they are further classified in accordance with their degree of generality. This further classification is required in understanding if a concept represents a generic situation (applying across many situations) or a specific situation (applying to one or a few situations). The terminology used to label a concept, its level in the hierarchy, the degree of abstraction and the generation of appropriate generalizations are determined in a continuous comparison of cases. Similarly, each concept is identified according to the category to which it belongs and according to its semantic relationship with other concepts in the hierarchy. This has led to the evolution of a taxonomy of free-form design. For the purpose of our research, the taxonomy is understood as the essential concepts representing the semantics of the free-form design domain. We will use this taxonomy as a hierarchical concept vocabulary (Figure 14.5) and to define the framework for organizing the knowledge content of free-form design.[2] The features, at the higher levels of the hierarchy, are context independent, such as 'surface representation', 'framing strategies' and 'architectonic expression'. At the lower levels, they start to define more specific and context-dependent concepts, methods, tasks or product features, such as 'Gaussian analysis', 'developability' and 'sectional contouring'.

2.1.4 The evolution of a knowledge framework

Within the taxonomy, we recognize two sets of relationships between concepts with regard to the knowledge content they represent (Figure 14.6). One is the hierarchical relationships between the concepts in each category, providing an outline of the tasks and processes of the domain under study. The second relationship is the associations between the concepts within and across categories. Accordingly, both the concepts and the links between them can be considered as the variables that vary in each case and that

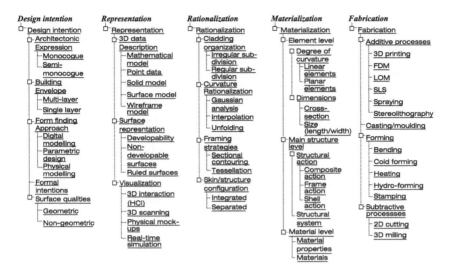

Figure 14.5 The taxonomy – a hierarchical concept vocabulary of digital free-form design.

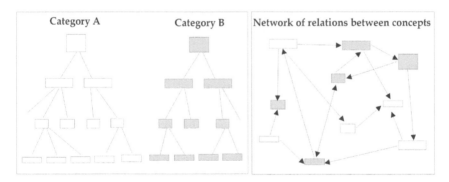

Figure 14.6 The types of relationships defined by the knowledge framework.

Left: Hierarchical relationships between concepts under each category. Right: A network of relationships between concepts within and across categories.

can be used to explain the unique ways in which designers frame design problems and generate solutions.

As was reported in section 2.1.2, the concept and its association with another concept or other concepts accommodate different knowledge types. While some concepts and their links refer to declarative and procedural knowledge, also referred to as process or task knowledge, some others are identified as situated and strategic knowledge elements, which may vary from one project to the next. Similarly, the links between these concepts may also change in meaning and form. It has also been observed that new concepts are generated within practice with the emergence of new techniques and methods in response to the specific needs of the designers. Therefore, rather than a static and formal description of the domain concepts, we propose to extend our taxonomy with the *theoretical model* developed throughout the horizontal analysis. This theoretical model consists of a set of propositions expressing the relationships between the concepts of the domain and knowledge construction. Thus, we propose a *knowledge framework* which consists of both the formal and the theoretical descriptions of the domain semantics. A knowledge framework further clarifies the knowledge structure of the domain, links the structuring of knowledge with its unique content, facilitates the assimilation of new concepts into the existing structure, and allows its users to share and extend their knowledge with others. Consequently, it provides the means for users to explicate their knowledge and supports the individual and collective construction of knowledge.

As for the knowledge content it is intended to support, we claim that the knowledge content of digital free-form design can be explained either as an instance of each concept or as an instance of the interconnection between two or more concepts in this framework. Therefore the framework should also allow extension, with the introduction of new concepts and links, which we define as knowledge construction.

2 The development of a web-based system to support collaborative knowledge construction

Based on the findings of the previous section, this section will describe the development and implementation of a web-based system (InDeS) to support collaborative knowledge construction, sharing and reuse. A prototype has been developed by implementing the knowledge framework into an existing database structure and by adding supplementary functionalities to its representational structure for efficient access to the related knowledge content. Using the characteristics of the domain content identified earlier, the prototype aims to provide a flexible structure for the organization and representation of the domain knowledge. While doing so, it tries to answer the following questions:

- How can the knowledge framework accommodate change and incorporate different design experiences and new information?
- How can it evolve through the actual creators of the knowledge themselves, thus contributing to a collective and collaborative creation of knowledge?
- How can this facilitate knowledge transfer between designers within and across disciplines?

Finally, the prototype is evaluated according to the factors which influence its effectiveness and further improvement with regard to user profiles in varying collaborative contexts.

3.1 A review of intelligent support for design

Artificial intelligence (AI) in design includes modelling of designer activity, the representation of designer knowledge, and the construction of systems that produce designs or systems to assist designers (Brown 1992). Knowledge-based systems rely on the extraction of generalizable and useful characteristics of the information and its classification in a manner that is retrievable as well as applicable in similar future design situations. Knowledge representation is considered as a prior condition to the development of knowledge support tools (Brodie *et al.* 1984). Case-based, knowledge-based, expert and other similar systems rely on the explicit symbolic representation of knowledge. These systems have mainly gone in two parallel directions in the support they provide for designers. First is the 'automated design systems', also called intelligent CAD (MacCallum 1990), whose aim is full or partial automation of the design process, while the role of the human designer is to give the initial requirements, evaluate solutions and build prototypes. The second is the 'design support systems' that aim at assisting human designers in their tasks by recalling past cases (Watson and Perera 1997), critiquing and navigating (Fischer 1992), and reasoning and consistency maintenance (Smithers *et al.* 1990; Tang 1997). A design knowledge support system, unlike a CAD system, does not actually design anything; instead, it attempts to support designers during the exploration of possible designs that could eventually help them to reflect on their design decisions, become familiar with the problem structures and possible solutions, and share their design knowledge. While the former approach provides a design memory with facilities that automatically retrieve or adapt previous cases, the latter provides a memory for indexing and retrieval of previous cases. In both approaches there is a strong need to develop a formal representation of the design experiences. In the following sections, we will report on the development of a web-based system (InDeS) which is based on the general principles of design support systems.

The extent of knowledge we intend to represent covers a broad spectrum of information necessary for the overall design and realization process, facilitating the exploration of collaborative knowledge in free-form design. It is anticipated that, through communication and collaboration, domain knowledge could be shared by all participants of the design team and contribute to the collective creation of collaborative knowledge. In this context, knowledge is explained as the meaningfully organized accumulation of information through experience, communication or inference (Zack 1999).

3.2 Highlighting the general features of design support systems

Various approaches and systems have been reviewed in the general domain of architectural design which provide knowledge support in the design process. Analogies between design systems can be based on various criteria. The identified similarities help us to highlight those features of design support systems that have a high degree of generality and can be applied to many systems.

Indexing is crucial in determining how the system will be used and how the cases will be retrieved, and provides a reasoning process for using the knowledge in the system by remembering the cases with common attributes to assist the user in comparing those cases with the problem at hand. The organizational structure of the memory contributes to the capability of accessing relevant knowledge (Oxman and Oxman 1990). Accordingly, a common issue in organizing design cases is the need to predetermine the features (abstract concepts) to serve as indices for efficient access to cases (Maher and Garza 1997) and to direct the users' attention to the related material (Domeshek and Kolodner 1992). A slightly different approach to memory organization is creating a memory structure as an associative network of stories related to these high-level concepts, resulting in a semantic net which provides the basis for indexing. In such schemes, instead of indexing the complete designs, every design story is indexed independently (Oxman 1994).

Representation of a specific case knowledge is another important issue. In the systems reviewed, it has been observed that case knowledge is represented either as complete representation of the entire case or in the form of knowledge chunks (also called design stories) comprising graphical, textual or numerical information. Memory organization will also affect the retrieval of domain-specific knowledge. For example, in typological models, instances and higher-level generalizations are linked hierarchically from general (type) to specific (case), whereas in precedent-based systems they are linked cross-contextually, which enables access to the generalized knowledge in a conceptual network.

The identification of the relevant features for the system is to a greater degree dependent on the knowledge content and will certainly affect the structure and organization of the memory and the representation scheme in relation to the context that is represented. In addition to the underlying cognitive model employed in various systems, the following aspects are what distinguishes each one of these systems:

- the scope and the domain;
- the content and the context;
- the design stages addressed;
- knowledge-encoding strategies.

In the following sections, how these variables are constructed and defined for the implementation of InDeS will be explained and discussed.

3.3 *The methodological framework*

The underlying cognitive model of InDeS is based on the knowledge framework developed earlier, as well as the theoretical standpoint adopted in our research. According to this standpoint, design is perceived not only as an information-processing activity, nor solely a problem-solving activity (which are the main foci of knowledge-based and case-based systems, respectively), but rather a combination of both. Furthermore, a very important aspect of the design process, especially in a new design domain where very little knowledge is present for the designers, is understanding the problem structures of the domain. Accordingly, supporting the designers not only in presenting earlier solutions to problems but also in providing them with the problem structures inherent in those solutions has become an important requirement for the system.

Another important requirement that influenced the functionality of the system was the extent to which it could reflect designers' actions and behaviours during design. It is commonly known that architects do not consider the different aspects of a design separately, but always in relation to other issues (Lawson 1997), and all the different considerations run through the heads of designers, simultaneously jumping from one design aspect to another. Accordingly, the system, while supporting the designers at the early stages of the design process, aims at providing information about the various stages of a project lifecycle. It is claimed that making evaluative material available to architects early in the design process can make them more aware of the downstream implications of their decisions. Consequently, it has been decided to organize the memory of InDeS in such a way that information pieces from the different stages of a design process can be related at various levels of detail. This could also allow the explication of problem structures in a multidisciplinary context. The main objectives for the implementation of InDeS can be summarized as follows:

- provides a conceptual understanding of the free-form design domain by clarifying problem structures in a multidisciplinary framework;
- a dynamic knowledge source which supports knowledge capture, creation, reuse and sharing;
- reflects human actions in design processes and can accommodate multiple views of knowledge.

3.3.1 *Knowledge content: type, quantity and acquisition method*

The determination of what design knowledge must be captured must be preceded by an understanding of the knowledge needs of the domain. Facing a new design problem, experienced designers do not design strictly by abstract design principles, nor do they exhaustively search a space of previous design cases. They refer both to previous experiences and to general domain knowledge. In order to support this process, the system should strive to record significant and meaningful concepts, categories and definitions (declarative knowledge), processes, actions and sequences of events (procedural knowledge), the rationale for actions or conclusions (explanatory knowledge), and circumstances and intentions (strategic and situational knowledge) within the domain of free-form design.

In the system described, these different knowledge types are fitted into a context that could facilitate access to the three different forms of knowledge that designers use during a design process: general domain knowledge (about the high-level concepts of the domain), specific domain knowledge (about the different states of the domain concepts and their variables) and specific case knowledge (about specific experiences in specific situations). InDeS proposes a representational scheme which integrates all three, which are stored as 'documents' in the system in various formats, such as text files, drawings, photographs and so on. The documents contain chunks of knowledge and are differentiated according to the context of use and the type of knowledge they contain. Instead of dealing with the entire design problem or process at once, the system is designed to focus on smaller pieces of information. Therefore the content of the documents should focus on this aspect by decomposing knowledge into relevant information pieces that can also be used in the indexing of these documents for ease of access to their knowledge content. The elementary tasks associated with the representation of documents are the content and the structure of the knowledge representation. The content needs to be identified in terms of what is in a document in order to reason about its applicability in a new design situation. If the repository is conceived as a 'knowledge platform', then many different views of the content may be derived from a particular repository structure (Zack 1999). Therefore it is helpful to provide a memory structure to define the contexts for interpreting the accumulated content. The structure refers to determining an appropriate structure of encoding design knowledge in a document. The content and the structure of a document determine how knowledge can be represented so as to maximize its usefulness for the user.

3.3.2 Knowledge context

During the design process, the architect has to investigate, evaluate and process a massive amount of cross-disciplinary information. In order to succeed, an understanding of the contexts in relation to the knowledge ingredients (content) is required. This refers to binding the pieces of information into some logical, contextual structure in order to provide answers on a general level to evolve a conceptual understanding of the domain under study. Consequently, the system should not only be designed for information retrieval but also be able to reveal the relations among the categories of information (Bar-On and Oxman 2002) according to the way in which information is used and manipulated by human designers.

The context for the system is created with regard to the knowledge framework that has been described in the previous sections. The framework provides a decomposition of particular design phases (five categories: design intention, representation, rationalization, materialization, fabrication) in the domain of digital free-form design. These five categories provide the context where the design system operates. These are the recognized categories of various design stages relevant to both design and realization processes, drawing architects' attention to all lifecycle implications of their design early in the design process. A further focus is placed on the design of the building skin(s) and the supporting structure.

3.4 Memory organization and representation

Memory organization refers to the way documents are organized for access during retrieval. If the memory does not contain many documents, then the memory organization is usually a list of pointers to the cases. The memory structure of InDeS is composed of two layers: 1) the concept layer and 2) the information layer. The concept layer consists of a hierarchical tree structure of concepts (described in section 2.1.3), which is the main representation scheme of the system (Figure 14.7).

The scheme consists of five generic classes of design phases (categories) with their associated features and sub-features representing a class of information elements hierarchically organized from more generic to more specific concepts. The information layer consists of documents that are connected to these features and to the links between the features. The representational medium that was used for the system is an extension of the database model developed for the InfoBase project, which was previously developed as a flexible representation framework that offers support for various information structures in different situations.[3] InDeS was developed by providing a context to this representational framework. This was achieved by adding a hierarchically organized vocabulary of features and additional functionalities for efficient access to the related knowledge content.

3.5 Document types in relation to features

Documents contain either domain knowledge (specific or general) or decomposed specific case knowledge in the form of knowledge or information chunks. Two document types are distinguished according to the number of features they are associated with and according to the type of knowledge they provide. Feature documents are associated with only one feature, and provide general and/or specific domain knowledge about different ways

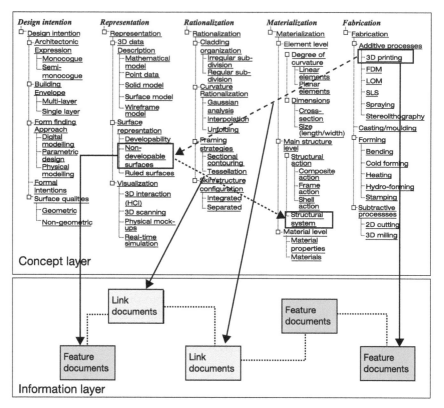

Figure 14.7 The connection between the information and concept layers.

that feature has been interpreted and used in a precedent situation. For example, a feature document connected to the 'non-developable surfaces' feature may contain specific or general information on the characteristics of non-developable surfaces, or may exemplify a method that was used to identify non-developable surface parts on a building model.

Many documents can be connected to the same feature. Link documents are associated with two or more features at any level in the hierarchy, within the same category or across categories. They are created by linking two or more features and may provide specific domain knowledge and/or specific case knowledge about how two or more features have been associated in a specific context. Link documents create conceptual associations between features. These associations define the problem structures within the domain which not only may reflect the objective side of a design situation but also may mirror the subjective preferences of the designer related to his/her own methodology. The link between the features fits the knowledge into a meaningful context. The links are augmented with distinct relationship types (e.g. constrained by, inspired by, dependent on, influenced by, defined by), mimicking the ways in which different pieces of information can be brought together by different designers in different design situations.

The system also allows the users to define new generic relationship types during document entry. These relationships are distinct not only in meaning but also in form (e.g. mono-directional, bi-directional, non-directional). Such differentiation of linkages represents the different viewpoints of the actors in the design team and further helps to explicate how the problems were structured by the designers that led to a specific solution. These solutions may be concrete or strategic, in either case providing tactical and conceptual support to the users. The linkages are another important aspect of cognition in design thinking. In associative reasoning, knowledge elements (features) are linked on the basis of these conceptual relationships. It is important to note that features under each category are intrinsically linked, and cross-categorical links between the features can be created only with document entry. This means that no links or associations exist between features by default. They are expected to be created and constructed in time by the users of the system.

3.6 Document descriptions for data entry and retrieval

A common issue in organizing the documents is the need to predetermine indices for efficient access to the documents. A common way of storing documents is by attribute-value pairs. This allows the document to be self-defining and incremental. An attribute (or property) identifies the attribute's value to a user. Associated with an attribute is a type (part of its meaning). In the system described, the documents are structured and identified according to their attribute types and values, and by the features they are associated with, each distinguished with a unique document ID. The types of documents are further distinguished by the number of features listed in their document descriptions, in other words by the number of features they connect (Figure 14.8). For example, a feature document is associated with only one feature, while a link document is associated with two or more features.

Consequently, features are cross-referenced via link document descriptions, creating a network of features which provide the means to navigate within the system by their associative connections. Searching within the documents is also possible via the descriptions of their particular attributes. While all documents have four searchable attributes (feature, project name, keyword, architect), an additional search function is provided for link documents via their link types. Link documents connecting only two features have an additional attribute called the link direction, which is not yet a searchable attribute in the current prototype. Non-searchable document attributes consist of comment, author and document type, which are listed in the information window during document retrieval for review purposes. All of these attributes can be defined by the user during document entry.

The system is designed for both reviewing and browsing through information, as well as adding new information into the system. The user interface is

Document type	Relations with features	Attribute types	Knowledge type
Feature document	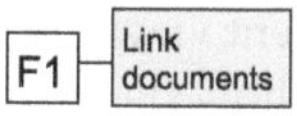 F1 — Link documents	Features, keywords, project, architect	'What' and 'how' knowledge, general domain knowledge, specific domain knowledge
Link documents (connecting two features)	 F1 — Link documents — F2	Features, keywords, project, architect link type, link attraction	'How' and 'what' knowledge, specific domain knowledge, case-specific knowledge, knowledge about problems, and solutions
Link documents (connecting more than two features)	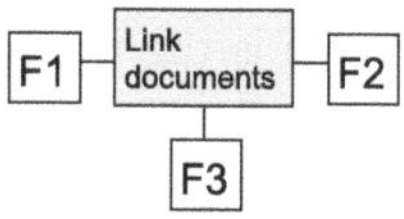 F1 — Link documents — F2 / F3	Features, keywords, project, architect link type	'How' and 'what' knowledge, specific domain knowledge, case-specific knowledge, knowledge about problems, and solutions

Figure 14.8 Comparisons between different document types according to the features they link, the attributes and the knowledge content.

designed to address these two purposes with separate interfaces. One is the 'browsing and search' interface (Figure 14.9). The other is the 'data entry and editing' interface (Figure 14.10), which allows the construction and extension of new knowledge into the system. The former is updated automatically according to the changes and additions in the latter.

Different documents may be entered into the system to exemplify how a feature is associated with another feature or features in a specific design situation, providing further information about the reasoning process and the viewpoint of the designer, linking his/her problem formulation to the solutions generated. Currently there are five default link types within the system (constrained by, defined by, dependent on, inspired by, influenced by), which have been specified during the testing of the prototype by using additional design cases from the free-form design domain.

The system is designed to allow the specification of new link types during document entry. The system is intended to support knowledge construction at any stage during the design process. As the design progresses, the designer

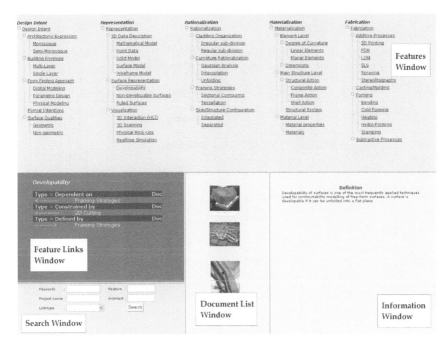

Figure 14.9 A snapshot of the 'browsing and search' interface.

learns more about possible problem and solution structures as new aspects of the situation become apparent and inconsistencies inherent in the formulation of the problem are revealed. Thus, during the design process, the designer often develops a better understanding of the particular problem he/she is working on. As a result, with new insights into the problem, a new view is formed and the problem and solution are redefined. This process of exploration and redefinition continues. This learning process may often result in new requirements being incorporated into the design specification, or may end up with the introduction of new information into the design knowledge base. The system allows a flexible and extensible framework for adding new features, link types and documents, as well as editing existing features and their definitions. However, how the knowledge content stored in the system will be used in the context of a new design situation and how it will be adapted and modified in a new context are totally up to the user.

3 Discussion on the applicability of the system in various contexts

InDeS is one of the various possible prototype implementations of a design support system facilitating knowledge construction, sharing and reuse in design. The contribution of this prototype lies in the fact that it is a first step

Figure 14.10 A snapshot of the 'data entry and editing' interface.

towards a concrete support for designers in digital free-form design which provides them with a situated and interdisciplinary view of design knowledge in their knowledge acquisition and utilization. The final prototype is not a 'complete system' and therefore has not yet been tested in different, real-life design settings.

In the domain of free-form design, with such a situated and interdisciplinary view of knowledge, the success and efficiency of InDeS, or any other system, will naturally be dependent on its ability to adapt to the changing contexts and to the varying profiles of users it intends to support. Therefore, a system which is based on knowledge construction, sharing and reuse in an interdisciplinary and dynamic context should also be evaluated according to the degree to which it serves this purpose in varying contexts. The varying contexts of use refer to the transfer of knowledge among users both horizontally and vertically. Horizontal transfer takes place among users with different disciplinary backgrounds or functional responsibilities, while vertical transfer takes place between users within the same discipline with different experience levels or different functional responsibilities. The following observations and evaluations have been made based on various discussions and interviews with relevant design experts and potential users:

- *Context 1: if the system is used between an architectural and an engineering office.* The system will be most beneficial if the offices already have some degree of collaboration. An efficient use of the system would be as 'an integrative framework' to facilitate communication, coordination and understanding over a large amount of complex information. A common concern was about members of each office wishing to aggregate and classify concepts in a different way (e.g. under a different category). As long as the descriptions of the associated features are clear, this would not cause any problems, since the semantic links between the features are created independently of the category under which they appear in the interface. Another concern for the practitioners has been the issue of intellectual property and who owns the rights of the knowledge that is entered in the system.

- *Context 2: if the system is used between the designers of an architectural office.* The most distinctive contribution of the system will be facilitating knowledge transfer between the experienced and the novice designers within the office. While the novice designers could benefit by learning from the design strategies, especially regarding the ways in which the experienced members of the design team formulate problems, the experienced designers could benefit from the explicit representation of their thoughts under such a dynamic knowledge structure. The contribution of InDeS to 'design learning' is expected to increase sequentially with more document entry into the system. The designers interviewed had all agreed on the increasing efficiency of the system with real-time maintenance but not on conducting such maintenance by all designers within the office. Issues of updating, maintenance, lack of enthusiasm to invest in long-term gain, and staff costs, and concerns related to intellectual property have been identified as decisive factors inhibiting the effective use and the applicability of the system in the practice context as was originally intended.

- *Context 3: if the system is used in an educational context.* It is anticipated that the immediate application of InDeS in an educational context, as a teaching and learning environment, would be more realistic and effective than its immediate use in practice. First, this anticipation is based on the fact that the control, update and maintenance of the knowledge content between the instructor and the student is an easier task compared to the hierarchical and organizational complexities of the design practice. Second, the system would require some degree of customization for each design or engineering firm, as well as an adjustment time for the users to get familiar with the terminology of the system. For both reasons, it is likely to take a longer time to see the actual contribution of the system in practice.

4 Recommendations for future work

In our particular research, the theoretical framework was generated by examining the limited number of cases in free-form design practice, albeit in depth. More empirical grounding and comparisons will sharpen and enrich the concepts developed here and yield more complex understanding of the phenomenon. First, it is necessary to investigate different contextual factors. We concentrated mainly on the technological and socio-organizational factors that affect the knowledge content and its acquisition and utilization. More contextual factors (e.g. cost, environment, company size and culture) need to be examined to ascertain whether the proposed concepts and framework are relevant in other situations. It is also necessary to test the system in the context of design practice, in complex projects involving multiple disciplines. This would also ascertain the level and degree of customization that would be required in each context. It is important to test the crucial balance between the degree of customization that should be allowed within the system and the degree of modification that would turn the system into a different tool from that originally intended. The ways in which this could be preserved are crucial.

In the practice context, an important future improvement within the system would require the inclusion of import and export facilities from/to CAD and CAE tools for instant data entry and retrieval of a specific phase and state of an earlier project. Both engineering and architectural firms keep a record of various versions of their drawings at different stages of the project, each of which has its associated knowledge and relevant representations in graphical, textual or any other relevant form. Allowing the transfer and organization of these documents by the system would facilitate the capture and reuse of the abandoned partial information of past designs. It is possible that, by extending the dimensions and contextual factors, a more finely regulated classification system will be needed, for example to distinguish between the levels of detail in the definition of abstractions (features). This would require a more elaborated search and retrieval mechanism for the system proposed. Similarly, the links and associations between the features could be assigned weights for conflicting and multiple dependency situations. Integrating these into the system would eventually require much more sophisticated programming skills and techniques.

Acknowledgements

The author would like to acknowledge Dr Rudi Stouffs, for providing the InfoBase database model derived from the ICTO-InfoBase project developed at Delft University of Technology, Faculty of Architecture, Design Informatics Department; and Seyfi Bal, for his contributions to programming and developing the new database model of InDeS.

Notes

1 The data in Table 14.1 is indicative rather than final. New concepts have been regenerated and renamed recursively, by continuous comparisons between higher- and lower-level concepts extracted across the categories.

2 It should also be noted that, in addition to the concepts extracted from the cases studied, we referred to various literature (e.g. quoted materials from interviews, field notes) to check the consistency of the evolving vocabulary.
3 The database model is derived from the ICTO-InfoBase project, which was developed by the Design Informatics Department of the Faculty of Architecture at Delft University of Technology, led by Dr Rudi Stouffs (Stouffs *et al.* 2004).

References

Bar-On, D. and Oxman, R. E. (2002). 'Context over Content: ICPD, a Conceptual Schema for the Building Technology Domain', *Automation in Construction*, 11(4), p. 467.

Brodie, M. L., Mylopoulos, J. and Schmidt, J. W. (1984). *On Conceptual Modelling: Perspectives from Artificial Intelligence, Databases, and Programming Languages*. New York: Springer.

Brown, D. C. (1992). 'Design', in S. Shapiro (ed.), *The Encyclopedia of Artificial Intelligence*, 2nd edn, pp. 331–339. New York: John Wiley & Sons.

Domeshek, E. A. and Kolodner, J. L. (1992). 'A Case-Based Design Aid for Architecture', in J. S. Gero (ed.), *Artificial Intelligence in Design*, pp. 497–516. Dordrecht: Kluwer Academic.

Eastman, C. M., McCracken, W. M. and Newstetter, W. C. (2001). *Design Knowing and Learning Cognition in Design Education*. Amsterdam: Elsevier.

Fischer, G. (1992). 'Domain-Oriented Design Environments', Proceedings of the 7th Knowledge-Based Software Engineering Conference, IEEE Computer Society.

Jong, T. de and Ferguson-Hessler, M. G. M. (1986). 'Cognitive Structures of Good and Poor Novice Problem Solvers in Physics', *Journal of Educational Psychology*, 78(2), pp. 279–288.

Kocaturk, T. (2006). 'Modelling Collaborative Knowledge in Digital Free-Form Design', Ph.D. thesis, Delft University of Technology, Faculty of Architecture, Department of Building Technology.

Lawson, B. (1980). *How Designers Think*. Westfield, NJ: Architectural Press, Eastview Editions.

Lawson, B. (1997). *How Designers Think: The Design Process Demystified*, completely rev. 3rd edn. Oxford and Boston, MA: Architectural Press.

MacCallum, K. J. (1990). 'Does Intelligent CAD Exist?', *Artificial Intelligence in Engineering*, 5(2), pp. 55–64.

Maher, M. L. and Garza, A. G. D. S. (1997). 'AI in Design: Case-Based Reasoning in Design', *IEEE Expert: Intelligent Systems and Their Applications*, 12(2), pp. 34–41.

Oxman, R. E. (1994). 'Precedents in Design: A Computational Model for the Organization of Precedent Knowledge', *Design Studies*, 15(2), pp. 141–157.

Oxman, R. E. and Oxman, R. (1990). 'Computability of Architectural Knowledge', in M. McCullough, W. J. Mitchell and P. Purcell (eds), *The Electronic Design Studio: Architectural Knowledge and Media in the Computer-Era*, pp. 171–187. Cambridge, MA: MIT Press.

Smithers, T., Conkie, A. and Doheny, J. (1990). 'Design as Intelligent Behaviour: An AI in Design Research Programme', *Artificial Intelligence in Engineering*, 5(2), pp. 78–109.

Stouffs, R., Kooistra, J. and Tuncer, B. (2004). 'Metadata as a Means for

Correspondence on Digital Media', *ITcon*, 9, Special Issue Digital Media Libraries, pp. 129–142.

Tang, M. X. (1997). 'A Knowledge-Based Architecture for Intelligent Design Support', *Knowledge Engineering Review*, 12(4), pp. 387–406.

Watson, I. and Perera, S. (1997). 'Case-Based Design: A Review and Analysis of Building Design Applications', *Artificial Intelligence for Engineering, Design, Analysis and Manufacturing*, 11, pp. 59–87.

Zack, M. (1999). 'Managing Codified Knowledge', *Sloan Management Review*, 40(4), pp. 45–58.

15 The use of 3D computer visualisation methods in value management briefing and design studies

The case for rapid prototyping and the impact on industry structure

Steven Male

Overview

Value management (VM) is a methodological management style for managing value in projects. With it origins in US manufacturing, it has developed and become more widely adopted in construction. North American thinking dominated development in VM for the first four decades, whilst recent developments in principally Europe, Australasia and China (notably Hong Kong) have also seen divergence of thinking emerge, reflected in different published national standards. What is clear is that value management studies are conducted at discrete intervention points, with associated characteristic study styles using tailored methods and a number of tools and techniques built around workshop formats at these intervention points. The use of technology within the workshop process tends to be limited, revolving normally around presentations and to a lesser extent group decision support tools. Experience of conducting over 150 VM studies has identified one critical issue – the need to use 3D visualisation techniques in a range of buildings-related studies where team members during a workshop process need to understand the interactions between the use of space by an organisation and the associated briefing and design options. This requirement is particularly acute at the early stages where VM is being used either proactively to develop a design brief or potential design options or reactively when an existing design brief or design solution is being reviewed. The chapter presents case study vignettes where this requirement has arisen and discusses why it has arisen, and identifies the potential use of 3D computer visualisation methods in such situations, together with the implications that stem from this for the VM and value engineering (VE) method, the design process and supply chain members under different procurement routes. The chapter also presents a series of generic VM study styles where ICT visualisation technology could be of benefit and the consequences for different supply chain strategies. The chapter concludes with the argument that there is potential for the use of ICT visualisation technologies in VM and that this would enable rapid prototyping in such situations.

Introduction

VM, with its origin in the United States (Miles 1972, 1989; O'Brien 1976; Fallon 1980; Zimmerman and Hart 1982; Parker 1985; Dell'Isola 1988; Kaufman 1990; Mudge 1990), derives its power from being a function-oriented, team-based, process-driven methodology adopted at various stages of a project, and especially in the early design stages to analyse space requirements. Kelly and Male in the UK followed a similar trajectory to Roy Barton in Australia, contextualising VM into a UK national context, through consolidating their work from the mid-1980s on North American VE (Kelly and Male 1993; Male *et al.* 1998a, 1998b; Kelly *et al.* 2004). In South-East Asia, debate has also recently focused on the contextualisation of an appropriate VM methodology to suit that locality (Liu and Leung 2002; Shen and Chung 2002; Fong 2004; Cheah and Ting 2005; Wan 2007).

Developments in VM thinking and practice have resulted in a diversity of definitions, procedures and official standards internationally. Whilst SAVE International does not define the methodology per se, its standard uses the term 'value methodology', highlights that VM includes the processes known as value analysis, value engineering, value management, value control, value improvement and value assurance, and adopts an approved job plan and a body of knowledge (SAVE 1998, 2006). In this respect the standard is all embracing but lacks the wider managerial definition of value management emerging from Europe, Australia and New Zealand. The European Standard for Value Management (BS EN 12973 2000) defines VM as a style of management, whilst recognising that other methods and management techniques also based upon the concepts of value and function have developed, such as design to cost and functional performance specification. The standard indicates that the goal of value management is to reconcile differences in view between stakeholders and internal and external customers as to what constitutes value.

The Australian and New Zealand VM standard (AS/NZS 4183 1994) defines it as a structured, systematic and analytical process which seeks to achieve value for money by providing all the necessary functions at the lowest total cost consistent with required levels of quality and performance. The VM process is seen as centred around a participatory workshop involving a multi-disciplinary, representative group of people working together to seek the best value solution for a particular situation. The new Standard DR 04443 (Standards Australia 2004, 2006) modified the definition, indicating that VM is a structured and analytical group process seeking to establish and improve value and, where appropriate, value for money in products, processes, services, organisations and systems. It also acknowledges that value for money is closely associated with the more traditional applications of value analysis and value engineering in activities such as the design, procurement, operation and disposal of entities. Again, the standard sees the VM study process as centred on a participatory multi-disciplinary workshop and now uses the term 'work plan' instead of the 'job plan' originally used by Miles (1989) and adopted in the Australian and New Zealand 1994 version.

To conclude, this chapter takes the view that VM is a style of management, and as a methodology its objective is to reconcile differences in view between stakeholders and internal and external customers around what constitutes value. This is achieved through a managed, structured, systematic and function-oriented process involving a multi-disciplinary team, normally brought together within a workshop situation. The methodology has been applied to products, services, projects, programmes of projects and administrative procedures.

Value management as a management style

It has been argued that function analysis is *the* only distinguishing characteristic of value management from other management philosophies and approaches. The European VM standard clearly indicates this is increasingly open to challenge. Additionally, many of the tools and techniques used by value managers are not unique to the methodology: holding workshops is not unique and neither are facilitating teams; there are also many management methodologies that have a structured process to them.

Value management has been placed within the context of projects, and a number of authors have developed and refined the concept of the project as a value chain, with its origins deep within the client organisation, relating value management interventions with programme and project management processes and the appropriate choice of procurement strategies (Bell 1994; Standing 2000; Male and Mitrovic 2005). The ideas behind the project value chain are set out in Figure 15.1.

Kelly and Male (1993) identified four levels across the project value chain where a VM study may intervene in a project delivery process, detailed in Figure 15.2. These levels have a direct impact on the type, purpose and deliverables from a value management study. The levels are:

- *Level 1:* concept (strategic project purpose or task) – the objectives and requirements of the investment decision or strategy;
- *Level 2:* spaces – space requirements to reflect organisational use;
- *Level 3:* elements – major building elements; the authors use the RICS BICS elements for this purpose and group them by a series of critical building functions related to Level 1;
- *Level 4:* components – major building components.

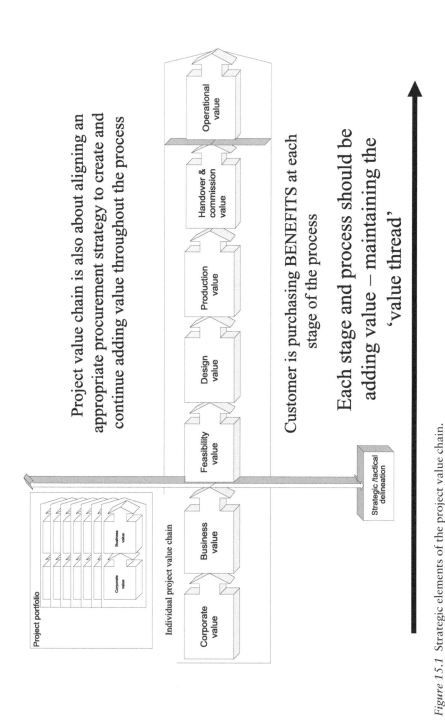

Figure 15.1 Strategic elements of the project value chain.

Source: Adapted from Standing (2000).

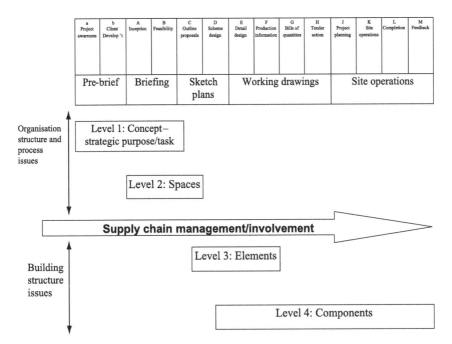

Figure 15.2 VM study levels.

Within the project value chain, each of the different levels results in a particular study style, with its own associated duration, methodology, phases, team composition during workshops and also outcomes. Hence, a study style is defined as an outcome of the stage in the project life cycle at which a value study is carried out, the manner in which the process is conducted and the anticipated deliverables. The VM benchmarking study research (Male *et al.* 1998a, 1998b) identified six probable points in the project life cycle where value studies are beneficial and where studies are most commonly found in practice:

1   strategic briefing study;
2   project briefing study;
3   charette (C) – undertaken in the place of the studies at points 1, 2 and 4;
4   concept design workshop;
5   detail design workshop;
6   operations workshop.

VM as a mode or style of management is seen as having three primary core elements (Kelly *et al.* 2004). First, it includes a value system or systems that interact and that need to be made explicit and aligned or realigned in order for value-for-money decisions to be made appropriately. Second, it has an associated team-based process. Third, it incorporates the use of function analysis to promote a greater and deeper understanding across a value system or systems of why something is or is not required. These primary core elements interact through three generic phases that have been identified for VM studies (Kelly *et al.* 2004):

- *The orientation and diagnostic phase*, involving value manager(s) and value teams in preparing for the study, meeting with the commissioning client, project sponsor, and key stakeholders involved in the study, reviewing documents and possibly conducting interviews and briefings. It may also include understanding and structuring the value problem or challenges in detail, exploring possible competing value challenges, discussing promising solutions and exploring the way forward on completion of any subsequent workshop phase. Agendas for the workshop phase will be developed and the method and manner in which the workshop is to be conducted will be ascertained, including the choice of tools and techniques and their appropriate ordering. The implications for implementation of options and solutions developed from the workshop phase will also be considered.
- *The workshop phase*, involving a forum or fora where alternative and/or complementary stakeholder and multi-disciplinary perspectives on the value problem are brought together to explore and reach a way forward. It is the phase at which optioneering takes place, normally captured in a study report including an action plan to ensure that value solutions and options will be implemented in the post-workshop phase.
- *The implementation phase*, recognised in the international benchmarking study as one of the key areas where value management falls down, and an appropriate implementation strategy discussed with commissioning clients during the orientation and diagnosis phase. Wherever possible those responsible for implementation will be interviewed and identified in the action plan at the close of the workshop phase.

Equally, whilst each of the VM project-level study styles noted above has its own characteristics, the nature of the procurement strategy chosen will also determine who is involved in each of these types of study from within the supply chain and the implications for the VM methodology adopted.

Figure 15.3 presents the benchmarked intervention points for value management studies at stages of the project life cycle – termed here 'value opportunity points'.

This chapter will discuss the implications for visualisation technology and rapid prototyping within the context of value opportunity points 1 to 4 in Figure 15.3; each opportunity point will be discussed in turn and presented in the order 1, 2, 4 and 3C to ease clarity.

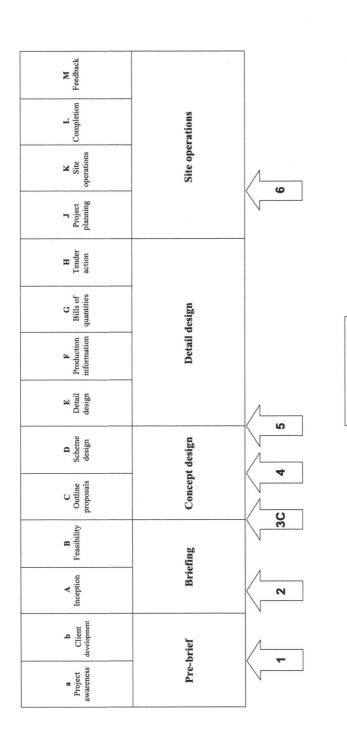

Figure 15.3 The project value opportunity points.

Source: Male *et al.* (1998a, 1998b).

The strategic briefing study – Project Value Opportunity Study Style 1 (PVO SS1)

The strategic briefing study is a Level 1 study noted in Figure 15.2 and is concerned with identifying the broad scope and purpose of the project, and its important parameters – articulating strategic needs and wants and the role and purpose of an investment decision as a 'business project' for the client organisation and clearly expressing the reason for that investment.

This study style answers the questions: why invest, why invest now and for what purpose? A strategic briefing study describes clearly and objectively the 'mission of the business project' and its strategic fit with the corporate aims of the client organisation. These corporate aims are explicit in terms of commercial objectives and usually implicit in terms of cultural values. The client's value system should be overtly expressed as a part of the strategic brief.

An important deliverable is the output specification explaining clearly what is expected of the 'business project'; it will also include establishing the outline budget and programme. The strategic briefing study will explore a range of options for delivering the 'business project', one of which could be the creation, refurbishment or renewal of a physical asset or assets as a corporate resource. 'Optioneering' could also involve developing and investigating non-physical asset alternatives. The strategic briefing study allows the decision to build to proceed with confidence.

Indicative techniques during the orientation and diagnostic phase of the study will include:

- interviews;
- stakeholder mapping;
- document analysis;
- questionnaires;
- post-occupancy evaluation of a similar facility or of the facility under discussion in the case of refurbishment and adaptation projects;
- site tours.

Indicative techniques during the workshop phase of the study will include:

- presentations;
- issues analysis, grouping, themes and prioritising;
- client value system;
- stakeholder analysis;
- strategic time line;
- project driver analysis;
- time/cost/quality analysis;
- functional analysis;
- function logic diagram;
- brainstorming alternatives;

- evaluation and development;
- presentations from working groups;
- plan for implementation;
- preparation and circulation of a study report from which the strategic brief will be developed;
- action planning.

Deliverables in the strategic brief document will include:

- the mission statement for the business project – a clear statement of why investment should be made now and in a given location;
- the project context;
- the client's value system and particularly how the success of the project will be measured;
- organisational structures for project delivery;
- overall scope and purpose of the project;
- high-level risks;
- programme, including phasing;
- a global capital expenditure budget and any cashflow constraints;
- initial options for inclusion in a procurement strategy;
- targets and constraints on operating expenditure and other whole-life costs;
- an implementation plan, including the decision to build or factors to be considered in the decision to build.

The strategic brief should present why an organisation has decided to invest in a physical asset or assets and why no other strategic options should be pursued that might compete for the same investment resource at that time. In the context of the UK government Office of Government Commerce (OGC)'s gateway review process this study would be at Gateway 0 – strategic assessment.

The project briefing study – Project Value Opportunity Study Style 2 (PVO SS2)

The project briefing study is a Level 2 study noted in Figure 15.2 and focuses on delivering the 'technical project', that is, the construction industry's response to client requirements embedded in the strategic brief. The project brief translates the strategic brief into construction terms, specifying performance requirements for each of the major elements of the project. If it is a building project, this will include spatial relationships. An outline budget will also be confirmed if a strategic briefing study has been undertaken, or developed if not. Indicative VM techniques during the orientation and diagnostic phase of the study will include those as for project PVO SS1 above and include the additional techniques of:

- benchmarking information from similar projects;
- post-occupancy evaluation of a similar facility or of the facility under discussion in the case of refurbishment and adaptation projects.

Indicative techniques during the workshop phase of the study will include those noted in PVO SS1 and will include the additional techniques of:

- process flowcharting;
- functional space analysis;
- spatial adjacency analysis;
- SWOT undertaken on any existing designs.

The project briefing document will include:

- a summary of the relevant parts of the strategic briefing document;
- the aim of the design, including priorities for project objectives translated into design parameters;
- the functions and activities of the client, including the structure of the client organisation and the project structure for delivering the project;
- the site, including details of accessibility and planning;
- the size and configuration of the facilities;
- the skeletal project execution plan or update in the case of it being independently prepared by the project manager;
- key targets for quality, time and cost, including milestones for decisions;
- a method for assessing and managing risks and validating design proposals;
- the procurement process;
- environmental policy, including energy;
- outline specifications of general and specific areas, elements and components in output terms;
- a cost-centred budget for all aspects of the project, including all elements of the construction project;
- options for environmental delivery and control;
- servicing options and specification implications, e.g. security, deliveries, access, work place, etc.;
- key performance indicators for each stage of the project.

To conclude, the primary purpose of this study is to develop the 'technical project' that is to be delivered by the construction industry as its response and/or deliver the strategic brief. The project brief provides the basis on which design can proceed. In the context of the UK government OGC's gateway review process this study would follow Gateway 1 – business justification. It is during this study that technology-based visualisation techniques could be an exceptionally powerful enabler during VM studies.

The concept design study – Project Value Opportunity Study Style 4 (PVO SS4)

The concept design study normally involves a value review of the initial plans, elevations, sections, outline specifications and cost plan of the proposed built asset. It may combine elements of a Level 2 study with a Level 3 study and also elements of Level 4 from Figure 15.2. The study will focus on validating the concept design or assisting the further development of design options and improvements. It is assumed that the client has agreed the project brief, although this will be tested as part of the study process. A good starting point for considering a concept design study is that, for most projects, the design has reached the point of seeking detailed planning permission. Indicative techniques during the orientation and diagnostic phase of the study will include those identified in PVO SS2 and may also include:

- post-occupancy evaluation of a similar facility or of the facility under discussion in the case of refurbishment and adaptation projects;
- facilities walk-through.

Indicative techniques during the workshop phase of the study will include those identified in PVO SS2 and the following additional techniques:

- major element function analysis and diagramming.

The outputs for the concept design study will include:

- a statement of the direction of the design;
- the project execution plan or update in the case of it being independently prepared by the project manager;
- the procurement strategy and the options explored for this;
- key milestones;
- key performance indicators;
- important risks, including a risk management strategy;
- a detailed cost plan and a detailed budget;
- a schedule of activities;
- the site layout and access, including the identification of ground conditions and any planning constraints;
- dimensioned outline drawings and an outline specification for all systems.

In the context of the UK government OGC's gateway review process this study would be at Gateway 2 – procurement strategy. This is a useful point at which to take an objective view of the proposed procurement process, incorporating the reasons for the decision and the actions in the workshop report and action plan.

To conclude, on completion of the concept design study, the design team may develop further options identified during the study or continue with normal design development in the full knowledge that the team has explored fully the design development to date and confirmed its acceptability to the client. Again, it is also during this study that technology-based visualisation techniques can be exceptionally powerful.

The charette – Project Value Opportunity Study Style 3C (PVO SS3C)

The charette is a hybrid study combining Levels 1 through to 4 in Figure 15.2 and is an audit of the project brief, is often undertaken once the concept design is complete, and audits the concept design against the strategic brief and project brief. In North America this study is often referred to as being undertaken at 10 per cent design (Kelly and Male 1993; Male *et al.* 1998a, 1998b). The charette is commonly the first study undertaken on a project. It implies that the client has reached the decision to build, completed the project brief, appointed a design team and then undertaken a value management study. The study is wide-ranging and comprehensive and normally incorporates the previous three studies discussed above – PVO SS1 to 4. The study focuses on validating that the project brief and frequently the concept design conform with and fulfil the client's value system. A primary purpose of the charette is to ensure that the client value system is overtly described and understood.

Indicative techniques during the orientation and diagnostic phase of the study will include those highlighted in PVO SS1 to 4, and indicative techniques during the workshop phase of the study will also include those highlighted in PVO SS1 to 4. The outputs from a charette will be a combination of the deliverables identified above in studies 1 to 3. The author's experience is that the charette is a common type of value management exercise undertaken on construction projects where it is a single event and is typical of the case studies explored further below.

To conclude, on completion of the charette study, the client value system will have been made explicit, the project brief will have been validated, and any outline designs will have been audited against the client value system, strategic brief and project brief. The design team will develop further options identified during the study or continue with normal design development in the full knowledge that the team has fully explored the strategic and project briefs and confirmed their acceptability to the client. It is during this study that technology-based visualisation techniques are an exceptionally powerful addition to a VM study.

Typical spatial analysis techniques used in PVO SS2 to 4

This section describes a series of techniques operating at Level 2 in Figure 15.2 and often used in a multi-disciplinary workshop situation for understanding functional space requirements. All space within a building should perform a function; if it does not that space is wasted and cost is incurred and adds no or very limited value. For maximum efficiency each space should have the highest degree of usage consistent with its function; therefore it is necessary to ensure that patterns of usage are reflected within a project brief. Equally the designer needs to ensure that circulation space, defined as essential non-functional space, is kept to the minimum consistent with the requirements of organisational efficiency.

Furthermore, organisational efficiency implies that space should be configured in terms of the client's value system and the client's proposed organisational structure and process. It is vital at this stage to recognise that the very fact the client is undertaking a project means that its organisation is about to change, with consequent probable impact on space use. Hence, space at Level 2 within a building has two major dimensions:

- *strategic space* – those essential major space areas and volumes that best describe the significant activities that the organisation requires space to be used for now and in the future – architects often refer to this within massing diagrams;
- *tactical space* – the translation of strategic space requirements into tactical delivery through spatial adjacency and organisational flow relationships.

It is in the area of spatial analysis and relationships that visualisation technology and rapid prototyping have an important strategic advantage in early-stage VM studies from PVO SS1 to 4. The activities associated with the function–space techniques will be set out below and can be utilised in a number of ways in combination.

Activity 1: determine users

The first stage in functional space analysis is identifying all of the users of the building. Invariably this will be a longer list than at first anticipated. Below are listed some of the users of a fictitious law court project using experience from VM studies conducted on high court and magistrate court VM studies:

- judge;
- judge's clerk;
- social worker;
- chairman of children's panel;
- solicitors;

- civil litigants;
- administration staff;
- police;
- custodial accused;
- non-custodial accused;
- jurors;
- witnesses;
- press;
- general public.

In a crown court VM project study there were 32 different users identified – creating a rich and complex mix of space use requirements and interactions. Equally, these numbers are not untypical of hospital projects, and in one instance close to 50 different end-users of space were identified when taking account of medical, ancillary, support or administrative staff, emergency services and patients.

Activity 2: spatial and organisational process flowcharting

Each user from the list in activity 1 is studied in turn and a flowchart prepared of their use of space. The reality is that where there are large numbers of different users identified these will be divided into typically:

- types of regular users;
- ad hoc users;
- users with special needs;
- atypical users.

A selection of user groupings is used subsequently to make the exercise more manageable.

Each activity of the user's daily routine is explored using process flowcharting within the building; each activity is connected by arrows to the next activity, with the presumption that each activity will require space. Even the activity of entering the building will require an entrance lobby of some sort, and the activity of moving from one space to another indicates circulation space. Figure 15.4 illustrates simplistically the activities of a judge's daily routine.

Activity 3: space specification

Each activity in the flowcharting exercise undertaken by users will require space; that space will have the attributes of size, servicing (heating, lighting, ventilation, acoustics, etc.), quality, normally defined by fittings and furnishings, and finally the technology support required. Activity 3 is a necessary precursor to the eventual compilation of room data sheets. Much of the

Flowchart for a judge

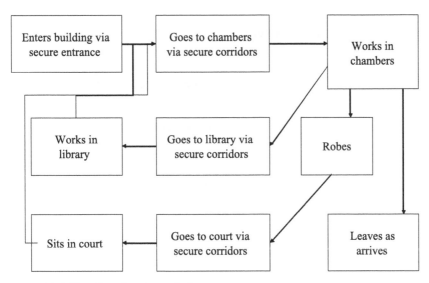

Figure 15.4 User flowcharting technique.

information contained within the space specification will be absorbed into room data sheets. On completion of activity 3 all of the required spaces to be contained within the structure should have been identified and their attributes understood. In a workshop situation this process needs to remain dynamic, and it is easy to get bogged down in the mire if spaces are considered one at a time from the flowcharts. It is more efficient to list all spaces from the flowchart and then group those spaces that are similar in terms of their attributes and function.

Activity 4: adjacency matrix

Each space is identified with a distinct name. These names are transferred to an adjacency matrix diagram (Figure 15.5). A scaling technique is used to illustrate the adjacency requirement on an index scale of +5 (spaces must be adjacent) to −5 (spaces should not be designed as adjacent). In this context adjacency means that there is a physical link between one space and another, normally a door or a short length of corridor.

Spaces with an adjacency index of 3 are normally within easy reach of one another, separated by for instance one flight of stairs or a reasonable length of corridor. Spaces with an adjacency index of 0 give the indication to the designer that the spaces have no adjacency importance one with another and therefore can be anywhere in relation to the total structure. Spaces with an adjacency index of −5 should be completely separated one from the other in

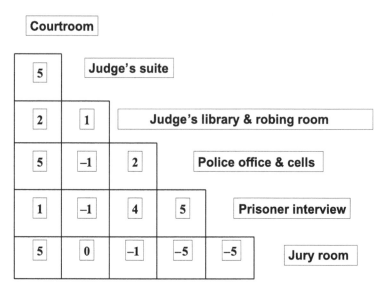

Figure 15.5 Adjacency matrix technique.

terms of environments, sound, acoustic properties and physical linkage. This does not mean that from a geometrical perspective the spaces cannot be separated by a single wall. However, the presumption is that one space can not be accessed from the other without travelling through many other spaces.

The adjacency matrix has been used in two situations: first, within buildings to define spatial relationships; second, between buildings in a master planning type of study.

Activity 5: space rationalisation prior to the preparation of room data sheets

A final study is undertaken of all spaces with similar services and environmental and/or function attributes. For example, in a study of one building it was determined that the conference centre and the employees' sports and social club both had the same structure and servicing requirements, both were two-storey height spaces and both required high levels of ventilation. The two spaces were therefore placed geometrically together, although under the rules of activity 4 both had an adjacency index of −5. The brief highlighted this situation but left the designers to ensure environmental and acoustic separation. In the final design the entrances to the two spaces were entirely remote, yet both spaces were able to share a dedicated plant room.

A space usage exercise is carried out to ensure that spaces with similar functional specifications are used to the highest degree. For example, the functional space analysis may highlight spaces of identical functional specification and yet they do not conflict on the organisation timetable. The client would need to decide whether these two spaces could indeed be combined. Finally, under this section and prior to introducing value engineering it should be stressed that the adjacency matrix and user flow can be a powerful audit tool during a value engineering exercise when analysing current designs.

To conclude, in building projects investment in strategic and tactical space and its pattern of usage involves significant cost. Equally, what is sought from that investment in space is significant benefits and added value. The five activities undertaken focus around balancing cost, benefits and the value expected from space use patterns over time. These techniques operate from a multi-dimensional perspective and normally require when using them that the value manager gets a team to create a mental picture of the facility that is being either briefed or reviewed; the effectiveness of this will depend on the skills and cognitive thought processes of individual team members. For technically oriented construction people this may be easier than for non-technical end-users. However, the essence of the techniques is to understand the complex relationship between space usage, spatial relationships, function, organisational processes and design options or solutions. Figure 15.6 captures the essence of the spatial analysis value challenge.

Technology has an important role to play within a VM workshop

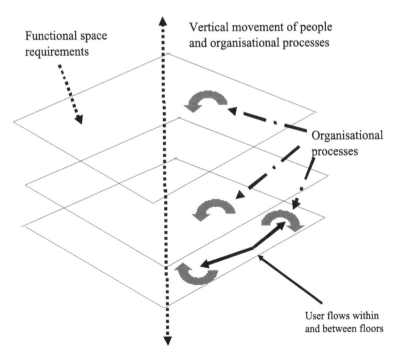

Figure 15.6 Functional space requirements.

environment with multiple stakeholders present and where these complex spatial relationships are being explored. The technological requirement is to be able to determine spatial configurations and parameters easily and quickly, explore the impact of organisational processes on special requirements and relationships, and equally be able to move around blocks of space easily and quickly such that all workshop participants can fully understand the emerging issues of any changes in organisational processes, spatial interactions and design changes. This becomes particularly important when organisational processes operate across multiple floors, are complex internally or also involve the use of multiple buildings. This technological requirement could be used in the generation of project briefs and concept designs as well as in the VM review of designs. Our experience in conducting over 150 VM and VE studies is that technology has been used in that way in only one instance – the case study presented below for master planning. In the majority of cases we have resorted to design drawings, sketches and concept designs or models during workshops, not computer-generated artefacts.

The use of computer-based technologically driven space analysis within a VM workshop could permit subsequently the option of rapid prototyping of designs from the outset or within review processes. The implications are, however, substantial. The issues will be explored through a series of small case study vignettes from VM studies conducted by the author.

Case study vignettes

The case vignettes have been chosen to represent the different approaches of using VM in the early stages of projects, where different methods have been adopted for conducting the studies. Each vignette draws out the implications of using the above techniques during the multi-disciplinary workshop stage.

A £16 million flagship library project and new £10 million science laboratory building for the same client

This library VM study for an atypical library was a charette, combining Levels 1 and 2 in Figure 15.2 with coverage of PVO Study Styles 1 to 4. It was undertaken over a number of weeks, culminating in a two-and-a-half-day workshop. The study, led by the author, involved a briefing team – comprising a full 'shadow design team' plus contractor, mainly for the workshop phase. The briefing architect within the team was involved with the author in the early stages of the study to confirm the strategic space requirements (essentially three main components forming the massing diagram): current library catalogue and periodicals usage and related study space, staff space and archive space. This was established through a dialogue with end-users and also a facilities walk-through of the existing library. The strategic space requirements identified were substantially less than originally proposed by library staff. This became subsequently an important focus of the workshop process, including establishing the related tactical space elements.

The VM workshop process involved taking the end-users and key client decision makers through a focused analysis of the project. The first half-day involved establishing the mission of the project, its drivers and budget parameters. The second day involved taking the library end-user team through a very structured analysis using a combination of user flow and spatial adjacency analysis within the context of the strategic space components. From this the full extent of spatial/function relationships was established. Day three involved taking the library end-users and key estate management client decision makers through a very structured elements and components function analysis to establish the parametric output specification and a budget range using benchmark data from similar projects. The end-users were placed under considerable pressure with a very full and comprehensive briefing interrogation by the design team. The third day also established the procurement strategy for the project with the assistance of the contractor member of the team. Subsequently, detailed data sheets were developed for inclusion within the brief. The result was a challenge to space assumptions and requirements, a sense from the library end-users that value had been compromised but not from senior client decision makers, an agreed procurement strategy and an agreed budget and control document for subsequent stages of the project.

The new science laboratory, for the same client body, was required, first, to build and test an experimental deep-space exploration space capsule housing experiments and, second, effect the co-location of another separate science department into the same building. The briefing architect and cost consultant who developed the project brief were the same as for the library, resulting in a combined PVO SS2 and 3 VM/VE-type study. In this instance the author was involved only to lead a one-day VE exercise, because the cost plan for the brief had indicated an emerging substantial over-budget problem, and end-users were not prepared to compromise on space and element function standards. Again, the end-user team, predominantly scientists, were taken through a very structured spatial analysis to test their space requirements in depth. The result was a challenge to these requirements and a sense from end-users that value had been compromised but again not from senior client decision makers.

Commentary: It was clear from both studies that, whilst the technical design/construction team had no problems envisaging the spatial and organisational functioning in combination, the end-users were having considerable difficulty and often related issues back to the current ways of operating rather than seeing the new buildings as having a direct impact on changing working patterns and also potentially offering opportunities to build new ways of working and associated organisational routines. The use of visualisation technology would have assisted greatly in both circumstances with end-users and would also have permitted rapid prototyping during the workshop process as spatial changes were discussed and agreed in a dynamic workshop environment. Whilst the corporate client was relatively knowledgeable in construction terms and was involved in both workshops, end-users were much less familiar with 3D methods of dynamic cognitive spatial thinking.

Master planning exercise – London urban regeneration scheme

The VM study was a master planning concept design and enabling infrastructure VM study, led by the author and working with the existing design team and other key stakeholders. The enabling infrastructure was costed in the order of £27 million, whilst the regeneration scheme was costed in the order of £150–200 million. The objective of the VM study was to confirm the master plan layout, obtain stakeholder buy-in for the master plan and subsequently finalise the impact on enabling works ready to go out to tender.

The architects had produced a visual walk-though and fly-through for the master plan of the whole site as the initial input into the workshop process. This educated the technically oriented as well as non-technical workshop attendees from the outset. The schematic walk-through presentation was also referred to during the ongoing workshop process at various stages and assisted greatly in the use of the adjacency matrix to agree the final layout of the footprints for the major building envelopes within the development. It also assisted in confirming, in conjunction with the adjacency matrix, the enabling infrastructure requirements prior to tender.

Commentary: This workshop demonstrated the power of the use of technology to assist visualisation of emerging issues.

VM and VE studies on two major court projects

These two studies represent the classical manner in which spatial relationships are tested in VM/VE studies where the value manager, in this case the author, was working alongside an existing design team but with a knowledgeable project sponsor also within the team.

The first case vignette is a £35 million flagship high court project, with a VE study essentially undertaken to verify the sketch design. Extensive use was made of spatial analysis, with adjustments being made to space relationships. The use of visualisation technology would have assisted in rapid prototyping of design changes, although the full team were competent in 3D visualisation using drawings.

The second case vignette is a £25 million magistrates' court requiring a reduction of £10 million in overall cost in order for the project to be sanctioned, representing major surgery on the existing design solution. Whilst other issues were also addressed, the team had to reduce considerably the spatial requirements of the project, and again the use of visualisation technology would have assisted in this dramatically.

Commentary: These two studies demonstrate that in an instance where space changes are made during the workshop process the full team would have been able to appreciate the ramifications of significant changes to spatial options using visualisation technology in such a dynamic environment. Equally, its use would have assisted in rapid prototyping of spatial solutions.

The impact of procurement strategies on the use of VM

The appropriate choice of procurement strategy links demand and supply sides together within a particular project demand and supply chain system (Male 2002, 2003b). The ease with which this system is formed, works and delivers an end product which is fit for purpose determines if value is added or reduced by the system overall. Schematically, this is set out in Figure 15.7 in terms of the project value chain (Kelly *et al.* 2004).

Within the project value chain, the client and main contractor are the primary protagonists in construction and, depending on the procurement choice adopted by the client and the design team, generate the major cost commitment for most projects. The latter is responsible for approximately 15 per cent of the client's project expenditure, primarily through design team fees. The main contractor is responsible for the major element of

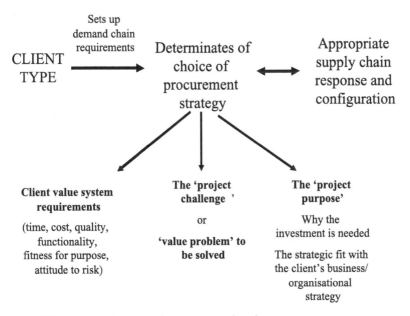

Figure 15.7 Strategic elements of the project value chain.

Source: Adapted from Obeng (1994); Hetland (2003); Male *et al.* (2004, 2007).

expenditure through the production process, some 85 per cent of the project cost during the construction phase.

Main contractors bring together and marshal an extensive supply chain network to meet a diverse range of client needs, and have the capability to manage multi-project supply chains. They operate with an organisational supply chain capability. Certain procurement routes, such as the traditional system, splitting design from construction, preclude leveraging the main contractor's knowledge, expertise and organisational supply chain capability much earlier in the project process to the benefit of the project and the client. The more recent procurement routes in the UK, such as prime contracting and the NHS Estates equivalent ProCure 21, are attempts at integrating supply chain members more closely with client representatives and leveraging the organisational supply chain capabilities of the main contractor. Depending on the procurement route adopted, the delivery of professional services can fall either within the early formative stages of the client's project supply chain, where they act as agents for and work directly to the client, or within the main contractor's supply chain, where they act as another skill set amongst many who work directly for the main contractor, perhaps only on a given project or across projects.

A programme of work involving the author has also postulated possible scenarios for the structure of the construction industry revolving around the requirements of the project value chain and supply chain management (Male and Mitrovic 1999; Mitrovic *et al.* 1999; Brown *et al.* 2000). The Elsewise project articulated that the core competencies of large-scale engineering (LSE) contractors are widening for the delivery of LSE projects. These are projects that are complex, multi-discipline engineering (design and construction) undertakings encountered at the top end of construction's activities. This type of project includes complex buildings, process plant, infrastructure and other significant civil engineering works. This work has been extended to include investigating the consequence of the above for the domestic UK construction industry using the structural steelwork supply chain as an exemplar (Male and Mitrovic 1999; Mitrovic and Male 1999; Mitrovic *et al.* 1999; Brown *et al.* 2000). Brown *et al.* (2000) proposed that construction supply chains in the UK would be much wider, operating regularly with different support structures and also with a greater reliance on information communications technology (ICT).

Equally, Holti *et al.* (2000) from their work on the Building Down Barriers project, argue that any approach which integrates supply chain members requires them to compete through superior underlying value by establishing long-term supplier relationships. Also, they envisage teams moving from project to project, with project activities integrated through clustering supply chain members to permit simultaneous engineering. Hill (2000), in a similar vein, argues that the principal objective of early involvement of the supply chain in construction projects is to manage cost effectively, improve functionality and as a consequence deliver better-value construction projects.

Effective organisational supply chain management capabilities for a main contractor will also require premeditated decisions on supply chain partners and strategic suppliers, perhaps those national/international firms from which they procure on a very regular basis or that are critical to certain types of project delivery. Tactical/operational supply chain decisions will be made from preferred and approved lists of suppliers for particular geographic locations. Equally, the organisational supply chain capability of the main contractor may need to be developed more fully to incorporate standardised procedures, processes and components, an increased focus on delivering value to met client requirements, the most appropriate organisation structural configuration to meet these demands, and greater use of information technology to coordinate and support collaborative working environments.

Typically, the primary procurement routes that now exist in UK construction are:

1   Traditional, with design split from construction, often now termed a non-integrated route. The use of VM is optional within this procurement route, and it can be overlaid with project-specific partnering philosophies to encourage greater levels of collaboration.
2   The management forms of procurement, construction management and management contracting, acting as overlays on the traditional route for more complex projects – again often now termed non-integrated routes. The use of VM is optional within these procurement routes, and they can be overlaid with project-specific partnering philosophies to encourage greater levels of collaboration.
3   Design and build (D&B), and its variants such as turnkey, package deal and early contractor involvement (ECI), often now termed an integrated route. The use of VM is optional within these procurement routes, although within ECI it is recommended as best practice by OGC and other government clients that adopt this approach. Both D&B and ECI can be overlaid with project-specific partnering philosophies to encourage greater levels of collaboration.
4   Prime contracting and its variants (e.g. NHS ProCure 21, alliances with technology clustering), termed an integrated route. VM is mandatory within these procurement routes.
5   PFI/PPP, termed an integrated route, although the evidence suggests that this could be a misnomer. The use of VM is optional within these procurement routes, although it is recommended as best practice by the OGC for all government clients.

Table 15.1, from Male and Mitrovic (2005), brings together client characteristics and the relative degree of use of non-integrated versus integrated routes. Non-integrated, partially integrated (using an overlay of project-specific partnering) and integrated procurement strategies refer to the extent to which the opportunity exists for a comprehensive project supply chain to

Table 15.1 Clients, and project value, procurement and supply chain

	PRIVATE SECTOR					PUBLIC SECTOR				
	KNOWLEDGEABLE — REGULAR PROCURERS			LESS KNOWLEDGEABLE — INFREQUENT PROCURERS		KNOWLEDGEABLE — REGULAR PROCURERS			RELATIVELY KNOWLEDGEABLE — RELATIVELY FREQUENT TO INFREQUENT PROCURERS	
Client Type / Project Value Chain Requirement	Consumer Clients: Large Owner Occupiers	Consumer Clients: Small Owner Occupier	Speculative Developers	Consumer Clients: Large Owner Occupier	Consumer Clients: Small Owner Occupier	Speculative Developers	Consumer Clients: Large Owner Occupier	Consumer Clients: Medium to Small Owner Occupier	Consumer Clients: Large Owner Occupier	Consumer Clients: Small Owner Occupier
Unique projects			NA	✓ Non Integrated		NA	✓ Integrated			
Customised Projects		✓ Non Integrated	✓ Non Integrated	✓ Non Integrated	✓ Non Integrated	NA		✓ Partially Integrated & Integrated	✓ Partially Integrated & Non Integrated	✓ Partially Integrated & Non Integrated
Process Projects	✓ Integrated	✓ Non Integrated				NA	✓ Integrated	✓ Partially Integrated & Integrated	NA	NA
Project Portfolios	✓ Integrated		✓ Integrated	NA	NA	NA	✓ Integrated		✓ Partially Integrated & Non Integrated	NA
PVC Orientation	*Sophisticated Leaders* Internal Advisers Use VM; likely to require use of ICT Visualisation Technology by Supply Chain	Followers External Advisors Unlikely to Use VM; unlikely to require use of ICT Visualisation Technology by Supply Chain	*Sophisticated Leaders* Internal Advisers Use VM; likely to require use of ICT Visualisation Technology by Supply Chain	Reluctant Followers External Advisers Wait and See Unlikely to Use VM; unlikely to require use of ICT Visualisation Technology by Supply Chain	Reluctant Followers External Advisers Wait and See Unlikely to Use VM; unlikely to require use of ICT Visualisation Technology by Supply Chain		*Sophisticated Leaders* Internal Advisers Use VM; likely to require use of ICT Visualisation Technology by Supply Chain	*Sophisticated Followers* Internal and External Advisers Use VM; likely to require use of ICT Visualisation Technology by Supply Chain once proven	Followers External Advisers Wait and See Unlikely to Use VM; unlikely to require use of ICT Visualisation Technology by Supply Chain	Reluctant Followers External Advisers Wait and See

Source: Male and Mitrovic (2005).

Note: The ✓ denotes that this is the probable occurrence, NA indicates no occurrence, and a blank indicates a possible but unlikely occurrence.

be either involved or not involved with the client from the outset of the project and throughout its delivery, into use and beyond.

In terms of client characteristics, knowledgeable, volume-procuring clients are in a position of considerable market power to set out their own project-focused project value chain requirements. Infrequent procurers, numerically greater in number, are in a much less powerful position owing to a much lower volume of ad hoc spends. It is well noted that these types of clients, if in the private sector and less knowledgeable, place a heavy reliance on their professional advisors, often slipping unwittingly into the traditional method of procurement where consultant architects or engineers have been approached first. In the public sector a range of different types of clients will exist, from the large corporate central government department or agency volume procurers to the medium to small local authorities, and to those autonomous organisations that are funded through central government funds, such as sponsored bodies or universities. Equally, they will have to adhere to (especially central departments) or be guided by government policy (especially local government and universities) on procurement regimes.

One of the key defining characteristics concerning level of client knowledge about construction industry operations is the extent to which it is embedded in the organisation because it is seen as essential to core business activity and is also directly related, as discussed above, to volume and frequency of procurement activity. Very knowledgeable clients are those that are high-volume procurers, that have a large physical asset base supporting the core business activity and where knowledge about construction is embedded deep within the organisation. Typical examples in the UK are the Ministry of Defence, the Highways Agency, the National Health Service, rail infrastructure operators, large airports, either public or private, and large speculative developers.

Relatively knowledgeable clients are those that require physical assets to support their core business, but the assets may not be perceived as central to organisational functioning since knowledge of construction is not embedded deep within the organisation. Examples include universities, which are relatively large and at certain times in their organisational development may have substantial building programmes and have estates departments. Less knowledgeable clients will be very infrequent procurers, where knowledge might reside only within a few people, if that, and with a heavy reliance on external advisors. A consolidated client typology is presented in Table 15.1, bringing together the impacts of different types of clients, level of supply chain integration, and project value requirements with the use of value management.

It should also be kept in mind when reading Table 15.1 that central government clients, whilst they have open to them a range of procurement strategies linked and tailored to the project value chain, now operate under the Office of Government Commerce 2002 guidelines and have only three procurement options recommended for use by them, the last three of the list above, notably design and build, prime contracting and PFI/PPP, recognised as providing value for money in the public sector. In line with this, it has been assumed that central and local government clients at all levels will experiment with increased levels of supply chain integration, through the use either of BDB initiatives (central government) or, at a minimum, of partnering approaches overlaid on other non-integrated routes (local government).

In conclusion, the impact of procurement routes directly affects the creation and delivery of value through the project value chain; some procurement routes assist this process; others may hinder it. In line with best practice recommendations from a number of industry reports, integrated approaches to procurement have VM as a mandatory element or its use is strongly promoted. The next section draws together the ideas proposed in respect of PVO Study Styles and VM generic study styles encountered in practice with the implications for the potential use of ICT visualisation technology within VM studies; subsequently this will be further incorporated into Table 15.1 and related to procurement strategies and client characteristics.

Generic VM study styles

Male *et al.* (2007) identified a series of generic study styles for value management, operationalised and informed by theoretical perspectives from business strategy, project management, the project value chain, and project programme management and their own involvement in a wider range of action research–consultancy activity. Figure 15.8 indicates the generic study styles.

Generic Study Style 1 (GSS1) occurs where an independent, appointed value manager works with any existing multi-disciplinary team using the structured VM process to assist the team to integrate and prioritise information, understand value problems, structure thinking and develop a way forward such that value systems are and remain in alignment. Challenging of assumptions is less in evidence, and the process, tools and techniques of value management are used to assist a team to think through a problem and find a way forward. It is fast and efficient, increases cross-team learning and information sharing, and focuses on solving the real problem at hand. The emphasis for the value manager in GSS1 will be more on the orientation and diagnosis and workshop phases and less on the implementation phase, since an existing team should usually be highly motivated towards implementation. This GSS has been adopted for assisting supply chain bid teams to develop a tender submission within the D&B procurement route, in project-specific partnering approaches or within tender situations for integrated collaborative procurement routes such as prime contracting. It has the potential to support the use of ICT visualisation technology in VM studies if the lead design member of the supply chain has such technology available. It is highly unlikely that the value manager will have access to such technology unless employed within a large consultancy organisation.

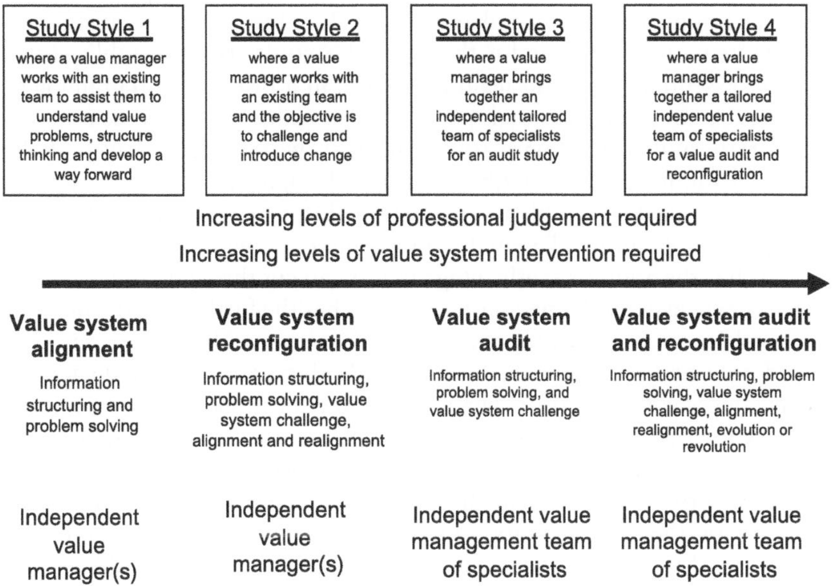

Study Style 1	Study Style 2	Study Style 3	Study Style 4
where a value manager works with an existing team to assist them to understand value problems, structure thinking and develop a way forward	where a value manager works with an existing team and the objective is to challenge and introduce change	where a value manager brings together an independent tailored team of specialists for an audit study	where a value manager brings together a tailored independent value team of specialists for a value audit and reconfiguration

Increasing levels of professional judgement required

Increasing levels of value system intervention required

⟶

Value system alignment	Value system reconfiguration	Value system audit	Value system audit and reconfiguration
Information structuring and problem solving	Information structuring, problem solving, value system challenge, alignment and realignment	Information structuring, problem solving, and value system challenge	Information structuring, problem solving, value system challenge, alignment, realignment, evolution or revolution
Independent value manager(s)	Independent value manager(s)	Independent value management team of specialists	Independent value management team of specialists

Figure 15.8 A study style continuum.

Source: Adapted from Male *et al.* (2005).

Generic Study Style 2 (GSS2) is the normal UK approach where an independent, appointed value manager works with an existing multi-disciplinary team with the objective of challenging and introducing change into the project or organisation's value system. It can be adopted for projects, programmes of projects, business processes or organisational change studies. It typifies the majority of the project-related studies conducted by the author and those identified in the international benchmarking study in Australia and in the private sector in North America. It is relatively inexpensive but places considerable reliance on the value manager being able to challenge the current perceptions, attitudes, assumptions and ideas of those in an existing team. This becomes more acute on complex projects, especially where there is a high degree of technological innovation or politics. As these increase or are found in combinations, a greater onus is placed on the value manager to achieve the appropriate balance of time between the orientation and diagnosis, workshop and implementation phases. This approach has been adopted across all procurement routes and on differing ranges of project complexity. Again, it has the potential to support the use of ICT visualisation technology in VM studies if the lead design member of the supply chain has such technology available. It is highly unlikely that the value manager will have access to such technology unless employed within a large consultancy organisation.

Generic Study Style 3 (GSS3) is where an independent, appointed value manager brings together an independent, tailored team of specialists for a value system audit study and is the classic US study style for public sector activity. Essentially an audit team intervenes in a project, a programme of projects or an organisational process. The value manager selects the appropriate multi-disciplinary team, and they work independently of the existing team. Its major strength comes from bringing together a fresh team to look at a problem, and provided the value manager selects the team appropriately it can handle any level of complexity within a problem situation. It has the additional advantage that it can cut through politics. However, it is expensive and time-consuming, and has the further disadvantage of increased learning curves for the independent team. The author has conducted only two such exercises in the UK; both were audits conducted for clients concerned about the progress and value for money of their projects. There is greater emphasis for the value manager on the orientation and diagnosis phase, less on the workshop phase, since he/she is much more in the role of process manager for his/her own selected team than challenger, and more on the implementation phase. The use of ICT visualisation technology in VM studies conducted as project audits is highly unlikely and unnecessary, since the emphasis is on examining what has been done and why on a project. However, the use of ICT visualisation technology in VM studies is potentially of use in a situation where an independent team has been employed to provide separate insights into a client's thinking; much will depend on the discipline base and organisational infrastructure from within which the value manager originates and his/her team.

Generic Study Style 4 (GSS4) is a hybrid between generic study styles 1, 2 and 3, where an appointed value manager brings together a tailored, independent value team of specialists for a value system audit and subsequent reconfiguration. Unlike the situation for GSS1 to GSS3, the value management team acts in an advisory role. The process is underpinned by the VM methodology; it is not bound by it in a rigid way, but uses its flexibility as a change-oriented process. The VM team works alongside client personnel to develop and implement solutions and be held accountable professionally for that involvement, the advice given and any recommendations made. The orientation and diagnosis phase with GSS4 could be extensive and workshops used in a variety of ways. Implementation is an ongoing process throughout. The author used GSS4 in the library case vignette noted above. The adopted method uses the VM process for gathering, sharing and exploring information, challenging assumptions usually but not always in a workshop, developing options and advising. The emphasis for the value manager with GSS4 will be on designing a reinforcing relationship between the distinct phases of the VM process, with orientation and diagnosis, workshop and implementation phases at times running concurrently as the change process unfolds. GSS4 will be designed specifically at times to go with the grain and at times against the grain. The use of ICT visualisation technology in VM studies is potentially of high use in this situation, where an independent team has been employed to work alongside and assist with a client's thinking; again much will depend on the discipline base and organisational infrastructure from within which the value manager originates and his/her team.

In conclusion, set against these different study styles, the role of the value manager becomes one of designing and implementing study styles to suit a whole range of different situations.

To conclude, earlier arguments have indicated the clear need to consider adopting and adapting ICT-based visualisation technology in the early stages of projects where function–space relationships are being worked through in VM studies; it could assist greatly the possibility for rapid prototyping – Table 15.2 indicates the likelihood of the use of certain study styles at different project intervention points and the opportunity for ICT usage within study styles. The analysis of generic VM study styles 1 to 4 has indicated not only greater value system intervention but a greater or lesser likelihood of using the appropriate type of technology. The key issue is the extent to which the design leader within a supply chain team has, or would be prepared to

Table 15.2 VM study styles and the likelihood of ICT implementation

		Generic VM Study Styles			
		VM Study Focus Information Sharing GSS1 Value System Alignment	VM Study Focus Challenging & Value System Reconfiguration GSS2	VM Study Focus Independent VM Team Value System Audit GSS3	VM Study Focus Value System Audit & Reconfiguration GSS4
Project Value Opportunity Points Study Styles	Strategic Briefing Study PVO SS1	✓ depends on procurement route	✓✓	✓	✓✓
	Project Briefing Study PVO SS2	✓ ICT probable depends on procurement route	✓✓ ICT probable	✓	✓✓ ICT probable
	Concept Design Study PVO SS4	✓ ICT probable depends on procurement route	✓✓ ICT probable	✓	✓
	Charette PVO SS3	✓ depends on procurement route	✓✓ ICT probable	✓	✓

Note: ✓✓ denotes highly likely; ✓ denotes likely

have, such technology in a VM workshop, or whether the discipline base and organisational infrastructure from within which the value manager originates and his/her team have access to such technology. Many VM study leaders – value managers – are sole practitioners and unlikely to have access to or invest in such technology; the costs would be prohibitive.

Discussion and ramifications

In the UK and USA value management usage appears to have attained a plateau, with signs that development is slowing or perhaps stagnating (Fong 2004). The reasons are that it is seen as lacking a professional image (79 per cent) and in decay (85 per cent), with the primary reason given for its decline being the ambiguous image (30 per cent). Fong also notes that the reasons given for the use of the methodology include cost reduction (24 per cent), performance improvement (14 per cent) and auditing (11 per cent). Cheah and Ting (2005) note similar findings from a South-East Asian perspective, indicating that, whilst 68 per cent of respondents were supportive of using VM/VE in construction, causes of its limited application included lack of support from those in authority (61 per cent), inflexibility in contractual provisions (61 per cent) and poor understanding about the methodology (59 per cent). There is also a perception that it is just a cost-cutting tool and that close alignments exist between the use of the term 'value' in VM/VE and its use in allied management techniques such as TQM. The most serious impediment to its further deployment is seen as lack of time to implement it on projects (65 per cent), although conflicts of interest (48 per cent) coupled with a lack of communication amongst stakeholders (43 per cent) and a divided/segmented project decision-making process (39 per cent) also came high on the list (Cheah and Ting 2005). In the UK, further development is also now being hampered by VM/VE being seen as closely aligned to workshop facilitation, which is only one of a number of key success factors identified by many researchers, writers and national standards dealing with VM. Also, in terms of process, the drive to shorter studies goes against the recommendations of Thiry (2002) on allowing sufficient time for these to occur, driving it increasingly towards a routinised process.

However, value management is an exemplar of good, structured, probing and challenging value-based problem solving. To undertake a study appropriately using a wide range of stakeholders takes time and thorough preparation to make it effective and efficient. The case argued here is that in the early stages of projects it has the capability through the use of ICT-based visualisation technologies to assist in rapid prototyping in a dynamic workshop situation where key stakeholders are present. Male *et al.* (2007) note that resolving conflicts of interest, increasing the levels of frank and open communication amongst stakeholders and reducing the impact of segmented project decision-making processes are fundamental reasons for adopting VM/VE as a more integrative process on any project regardless of procurement route adopted. When used within integrated collaborative procurement processes where it is mandatory it becomes a really powerful enabling technology for the methodology.

The consideration of the wider use of ICT visualisation technologies within VM studies undertaken within the early stages of projects has, however, many potential barriers within the VM workshop phase, not least that:

- it is costly to implement within a workshop setting;
- it may take additional time to work through effectively;
- the technology would require the support of fast internet access, often in non-office-based locations;
- VM study leaders have a tendency to be sole practitioners or operate within small consultancies or networks;
- it has the potential to change working practices throughout the supply chain if it becomes an accepted methodology for rapid prototyping;
- it is likely to remain within the domain of large-volume procuring clients and their supply chains, which in turn tend to be supply chain leaders that are themselves at the larger end of the organisational spectrum, often with large consultancy practices dominating their second-tier supplier groupings. This is addressed further below.

Industry restructuring through the use of ICT

The large main contractors, as principal organisational supply chain leaders, derive their competitive advantage from organising the procurement and management of their own multi-project supply networks. They are ideally placed to increase their market power and become supply chain leaders working directly for clients on a more regular basis. Their supply chain strategy, supported by ICT-driven collaborative enabling tools, becomes one of assembling the right teams, at the right time and in the right locations. This will drive them towards flexible, collaborative working methods, potentially involving them in considerable process design and redesign.

Research by the Warwick Manufacturing Group (1999) has indicated that many large clients are driving supply chain initiatives into the industry owing to their own, often global, competitive environment. Equally, there are purely domestically focused volume procuring clients exercising their market power. These are the clients advocating the Rethinking Construction agenda (Latham 1994; Egan 1998; NAO 2001, 2005; Strategic Forum for Construction 2002); they are committed to implementing collaborative ways of working throughout the supply chain, and are using their market power to drive this agenda forward. Government continues to support this move at all levels, although this is not without its critics (Green 1999). One obvious impact of the drive towards lean thinking (Ballard and Howell 1994; Moussa 2000; Gil *et al.* 2001) and collaborative working for the project value chain and supply chain management is the increased cost of up-front work on projects, acknowledged by both the BDB project and the National Audit Office report *Modernising Construction* (NAO 2001), the latter having put a cost of 1 to 1.5 per cent of project construction cost on this aspect for setting up collaborative team structures. Larger construction organisations, either contractors or consultants, with considerable financial power are unlikely to be deterred by this, since financial rewards and increased market power are likely to be greater. These will be the organisations investing in ICT enabling tools.

From a customer perspective, the winners are likely to be the large corporate clients that regularly use the skills of the industry. They are able to afford collaborative working methods. However, there is evidence recently in the construction and national press that the UK construction industry is beginning to complain about the cost of this type of working, with the increased bidding costs for prime and PFI projects, and equally with no guarantee of success or future workloads under Framework structures. If there remains a strong push towards the approaches discussed here, the losers are likely to be the small, irregular-procuring clients unable to afford the up-front costs, as indicated above, or to have the know-how to handle such approaches.

Similarly, some contractors, in terms of market power, have the advantage of organisational and economic size and are able to influence extensively their own multi-project supply chains owing to their purchasing power; others have much less capability in this area. In addition, a contractor's organisational supply chain strategy has to be attuned to the different procurement systems generated by client demand requirements. For example, some contractors have deliberately adopted a business strategy of exiting from traditionally tendered work and focusing their attention on increasing their workload through negotiated strategies, others remain committed to the former type of work, and others have moved towards PFI/PPP. Equally, large contractors have come to the fore in becoming supply chain leaders under prime contracting and its derivatives and in PPP/PFI. The large contractors and large consultant organisations are working closely together in a variety of consortium arrangements with large corporate, regular-procuring clients. The former's combined market power is increasing, and it is this group that will benefit from the use of rapid prototyping visualisation technologies within VM studies. The problem is that VM as currently practised is not set up to handle such developments, even within the integrated procurement routes such as prime contracting.

It is argued here that a different structure for the industry will emerge from that of today, with large corporate clients served by leading-edge supply chains, working collaboratively and using ICT-based enabling tools on a regular basis, with small irregular-procuring clients functioning, at worst, under traditional 'adversarial' approaches or at best under partially integrated procurement strategies. A key strategic issue for value management will be how it will adapt and respond to such changes – it has the potential through the use of ICT visualisation technologies in the early stages of projects to change and adopt different ways of working and create an appropriate environment for rapid prototyping.

Conclusions

Value management is a management style which focuses on value system evolution and resolution within projects by bringing the right team of stake-holders together at the right time. Through a structured, challenging, ana-lytical and often mediated process in the early stages of building projects it addresses the close alignment between the client's organisational function-ing and processes and the strategic and tactical use of space for building projects. A number of tools and techniques are used in the workshop phase of a VM study to address spatial and organisational relationships simul-taneously. It has been argued here that the potential use of ICT visualisation technologies would enable rapid prototyping in such situations. When used proactively in this manner technology-enabled VM has the capacity to align value systems from the outset and ensure that a project progresses effectively and efficiently and appropriate decisions are taken in light of the fact that it costs money to retrace footsteps. Space use is one of the major determinants of cost in building projects.

The power of the VM methodology derives from how a study is oper-ationalised. Structuring a study requires consideration of the value problem strategically and tactically, taking account of the different phases within the study, and the extent to which value systems (different stakeholder per-spectives) conflict or coalesce. In this context, the skills of the value manager will include the ability to understand a value problem, structure and manage the VM process, and bring different stakeholder representatives together. This chapter also argues that the role should be extended to include the appropriate choice of enabling technology at the right time. A number of barriers have been identified earlier, not least that often value managers are sole practitioners or small consultancies. The cost of such technology would be potentially prohibitive to them.

Through an exploration of the project value chain, procurement routes and an increased move towards collaborative forms of project structures, this chapter has also argued that the use of ICT-driven enabling visualisation technologies within VM studies is likely to be the domain of large-volume procuring clients, the large contractors and consultants. The chapter has also argued that this will change the structure and operation of the industry, in terms of both client service and service providers.

References

AS/NZS 4183 (1994). 'Value Management'. Joint Technical Committee OB6. Standards Australia and Standards New Zealand.
Ballard, G. and Howell, G. (1994). 'Implementing Lean Construction: Stabilizing

Work Flow', *Proceedings 2nd Annual Meeting of the International Group for Lean Construction*. Rotterdam, Netherlands: A. A. Balkema.

Bell, K. (1994). 'The Strategic Management of Projects to Enhance Value for Money for BAA plc', Unpublished Ph.D. thesis, Department of Building Engineering and Surveying, Heriot-Watt University.

Brown, D., Williams, P., Gordon, R. and Male, S. P. (2000). *Optimum Solutions for Multi-storey Steel Buildings*, Final report to Department of the Environment, Transport and the Regions.

BS EN 12973 (2000). 'Value Management', British Standards Institution.

Cheah, C. Y. J. and Ting, S. K. (2005). 'Appraisal of Value Engineering in Construction in South East Asia', *International Journal of Project Management*, 23, pp. 151–158.

Dell'Isola, A. (1988). *Value Engineering in the Construction Industry*, 3rd edn. Washington, DC: Smith, Hinchman & Grylls.

Egan, J. (1998). *Rethinking Construction* (Egan Report). Rotherham, UK: Department of the Environment, Transport and the Regions.

Fallon, C. (1980). *Value Analysis*, 2nd edn. Washington, DC: Lawrence D. Miles Value Foundation.

Fong, P. S.-W. (2004). 'A Critical Appraisal of Recent Advances and Future Directions in Value Management', *European Journal of Engineering Education*, 29(3), September, pp. 377–388.

Gil, N., Tommelein, I. D., Kirkendall, R. L. and Ballard, G. (2001). 'Leveraging Specialty-Contractor Knowledge in Design–Build Organizations', *Engineering, Construction and Architectural Management*, 8(5/6), pp. 355–367.

Green, S. D. (1999). 'The Missing Arguments of Lean Construction', *Construction Management and Economics*, 17(2), pp. 133–137.

Hetland, P. W. (2003). 'Uncertainty Management', Chapter 8 in N. J. Smith (ed.), *Appraisal, Risk and Uncertainty*, pp. 59–88. London: Thomas Telford.

Hill, R. M. (2000). *Integrating the Supply Chain: A Guide for Clients and Their Consultants*, Building Research Establishment Digest 450, August.

Holti, R., Nicolini, D. and Smalley, M. (2000). *The Handbook of Supply Chain Management: The Essentials*, CIRIA Publication C546. London: CIRIA and Tavistock Institute.

Kaufman, J. J. (1990). *Value Engineering for the Practitioner*, 3rd edn. Raleigh: North Carolina State Press.

Kelly, J. and Male, S. (1993). *Value Management in Design and Construction*. London: Spon.

Kelly, J., Male, S. and Graham, D. (2004). *Value Management of Construction Projects*. Oxford: Blackwell Science.

Latham, M. (1994). *Constructing the Team* (Latham Report). London: HMSO.

Liu, A. M. M. and Leung, M.-Y. (2002). 'Developing a Soft Value Management Model', *International Journal of Project Management*, 20, pp. 341–349.

Male, S. P. (2002). 'Supply Chain Management', in N. J. Smith (ed.), *Engineering Project Management*, 2nd edn, pp. 264–289. Oxford: Blackwell Science.

Male, S. P. (2003a). 'Supply Chain Management in Construction', *Proceedings of the Joint International Symposium of CIB Working Commissions: W55 Building Economics, W65 Organisation and Management of Construction, W107 Construction in Developing Countries*, Vol. 2, pp. 77–89, National University of Singapore, 22–24 October 2003.

Male, S. P. (2003b). 'Future Trends in Construction Procurement: Procuring and Managing Demand and Supply Chains in Construction', in D. Bower (ed.), *Procurement Management*. London: Thomas Telford.

Male, S. P. and Mitrovic, D. (1999). 'Trends in World Markets and the LSE Industry', *Engineering, Construction and Architectural Management*, 6(1), pp. 7–20.

Male, S. P. and Mitrovic, D. (2005). 'The Project Value Chain: Models for Procuring Supply Chains in Construction', Paper presented at Joint RICS Cobra 2005, CIB and AUBEA Conference, Queensland University of Technology, Brisbane, Queensland, Australia, 4–8 July 2005.

Male, S., Kelly, J., Fernie, S., Gronqvist, M. and Bowles, G. (1998a). *The Value Management Benchmark: A Good Practice Framework for Clients and Practitioners*, Report for EPSRC IMI contract. London: Thomas Telford.

Male, S., Kelly, J., Fernie, S., Gronqvist, M. and Bowles, G. (1998b). *The Value Management Benchmark: Research Results of an International Benchmarking Study*, Report for EPSRC IMI contract. London: Thomas Telford.

Male, S. P., Kelly, J. R., Gronqvist, M. and Graham, D. (2005). 'Re-appraising Value Methodologies in Construction for Achieving Best Value', Paper presented at Joint RICS Cobra 2005, CIB and AUBEA Conference, Queensland University of Technology, Brisbane, Queensland, Australia, 4–8 July 2005.

Male, S., Kelly, J. R., Gronqvist, M. and Graham, D. (2007). 'Managing Value as a Management Style for Projects', *International Journal of Project Management*, 25(2), pp. 107–114.

Miles, L. D. (1972). *Techniques of Value Analysis and Engineering*, 2nd edn. New York: McGraw-Hill.

Miles, L. D. (1989). *Techniques of Value Analysis and Engineering*, 3rd edn. Washington, DC: Lawrence D. Miles Value Foundation.

Mitrovic, D. and Male, S. P. (1999), *Pressure for Change in the Construction Steelwork Industry: Solutions and Future Scenarios*, Steel Construction Institute publication 293. Ascot: Steel Construction Institute.

Mitrovic, D., Male, S. P. and Hunter, I. (1998). 'The Impact of CIMsteel on Fabrication', *Structural Engineer*, 3 February, 76(3), pp. 50–54.

Mitrovic, D., Male, S., Hunter, I. and Watson, A. (1999). 'LSE Project Process and User Requirements', *Engineering Construction and Architectural Management*, 6(1), March, pp. 38–50.

Moussa, N. (2000). 'The Application of Lean Manufacturing Concepts to Construction: A Case Study of Airports as Large, Regular-Procuring, Private Clients', Unpublished Ph.D. thesis, University of Leeds.

Mudge, A. E. (1990). *Value Engineering: A Systematic Approach*. Pittsburgh, PA: J. Pohl Associates.

National Audit Office (NAO) (2001). *Modernising Construction*. London: NAO, Stationery Office.

NAO (2005). *Improving Public Services through Construction*. London: NAO, Stationery Office.

Obeng, E. (1994). *The Project Leader's Secret Handbook*. London: Financial Times Prentice Hall.

O'Brien, J. (1976). *Value Management in Design and Construction*. New York: McGraw-Hill.

Parker, D. E. (1985). *Value Engineering Theory*. New York: McGraw-Hill.

SAVE (1998). *Value Methodology Standard*. Dayton, OH: SAVE International.

SAVE (2006). *Value Methodology Standard*. Dayton, OH: SAVE International.

Shen, Q. and Chung, J. K. H. (2002). 'A Group Decision Support System for Value Management Studies in the Construction Industry', *International Journal of Project Management*, 20, pp. 247–252.

Standards Australia (2004, 2006). 'Value Management', DR 04443, Committee OB-006.

Standing, N. (2000). 'Value Engineering and the Contractor', Unpublished Ph.D. thesis, University of Leeds.

Strategic Forum for Construction (2002). *Accelerating Change*. London: Rethinking Construction.

Thiry, M. (2002). 'Combining Value and Project Management into an Effective Pro-gramme Management Model', *International Journal of Project Management*, 20, pp. 221–227.

Wan, Y. T. (2007). 'A Value Management Framework for Systematic Identification and Precise Representation of Client Requirements in the Briefing Process', Unpublished Ph.D. thesis, Department of Building and Real Estate, Hong Kong Polytechnic University.

Warwick Manufacturing Group (WMG) (1999). *Implementing Supply Chain Man-agement in Construction: Project Progress Report 1*, November. London: Department of the Environment, Transport and the Regions.

Zimmerman, L. W. and Hart, G. D. (1982). *Value Engineering: A Practical Approach for Owners, Designers and Contractors*. New York: Van Nostrand Reinhold.

16 Accelerating collaboration

Soft system imperatives for mobilising hard system developments

Mohan Kumaraswamy

Abstract

Rapid developments in hard system collaboration tools have overtaken current capacities of most construction organisations and personnel to effectively mobilise, let alone rapidly optimise, such multi-dimensional management systems. The usage and value of multi-criteria decision aids, rapid visualisation and prototyping tools and multi-agent operational systems, for example, are dependent on identifying and generating the required cultures, social dynamics and cooperative relationships amongst diverse and scattered project participants, often with divergent basic objectives.

This chapter draws on both international and local initiatives towards redressing the present imbalance between hard and soft systems, and addressing the growing gap in their future development. For example, the CIB Working Group W112 on 'Culture in Construction' and the recent Joint International Construction Research Workshop on the 'Social Dynamics of Collaboration in Projects' provide international pointers and reveal some general thrusts in this direction. These are reinforced by examples and experiences from Hong Kong-based research projects to develop 1) more inclusive and collaborative 'briefing' teams at the project front end, 2) 'management support systems' for large construction clients, 3) 'information and knowledge management systems' for contractors, and 4) 'relationally integrated value networks' for enabling continuous development. Barriers that decelerate, if not obstruct, the translation of such research and development into beneficial practice are noted. These barriers appear to grow from the mismatch with, if not distortion or underdevelopment of, prevalent industry cultures and social dynamics, and hence attitudes and relationships. A broader view of technology is advocated, in highlighting the value of the vital soft systems that must complement the hard systems. For example, the need to boost our lagging 'humanware' is noted, in order to effectively apply and optimise our rapidly accelerating 'technoware' within a holistic technological framework.

Introduction

The announcement of the 'Collaborative Human Futures' Workshop held in November 2007 noted that the increasing use of many technologies

> presents a new way of working that will not only change the way that designers work but also their function and role. It also changes the design and construction process and how clients procure their facilities. This way of working will not only offer opportunities to work collaboratively it will demand organisations to engage in new business partnerships and relationships.

This chapter responds to the above demand by mapping directions for improving partnerships, for integrating relational soft systems with each other, and for synergising them with hard technological systems. In a recent conference keynote address, Flanagan (2007) stressed the importance of people and the need for incentivising them to reflect their contribution to the design and production processes. He also cited research (by Baldwin, Thorpe and Carter; and Davidson and Moshini) that indicated construction cost reductions of as much as 25 per cent if information is transferred effectively.

However, such improvements are still not being achieved. For example, high-powered pushes for integration are falling far short of targets (Constructing Excellence 2006). Some of the barriers restricting such advances have been identified before, but not overcome as yet. For example, Erdogan *et al.* (2006): 1) repeated a 1999 observation by Betts and Smith in reminding us that 'the problem in the construction sector is not a lack of technology but more a lack of awareness of how to fully exploit it and how important culture changes are in order to allow this happen'; and 2) cited other researchers in cautioning that

> focusing too much on technical factors will result in technically excellent systems which are incompatible with the organisation's structure, culture and goals, since it neglects to consider how the new technology interacts with working practices, work organisation and job design, and work processes.

Taking note of the above, this chapter brings together snapshots of initiatives not merely to improve soft systems by themselves, as in the 'relational integration' examples, but also eventually to integrate them with harder technological systems as in the 'management support systems' and 'technology exchange' examples. The chapter concludes by noting the need for a combined onslaught against previously identified barriers by a battery of consolidated initiatives. For example, the slow uptake of an independent information and knowledge management initiative is noted, where not actively championed by large clients. The following sections are necessarily broad-brush summaries, given the wide canvas to be traversed in portraying relevant issues in this chapter.

Stakeholders, social dynamics and project cultures

Broadened scope of construction project management

Increasingly complex infrastructure megaprojects impact heavily on communities and the environment, for example those for transportation, hydropower, irrigation, water supply and sanitation. Those who research into, teach and/or practise 'project management' have hitherto focused on 'project execution', as lamented by Morris (2006), who points to such shortcomings in the US-based PMI body of knowledge and the 1998 Egan Report in the UK as examples. He calls for more emphasis on the front-end definitional stage of projects, for capturing and conceptualising project management knowledge in terms of it being a social as well as a technical process.

Similar thrusts emerge from elsewhere, for example the special issue on 'Managing Projects as Complex Social Settings' (Bresnen *et al.* 2005), the Workshop on 'Social Dynamics of Collaboration in Projects' involving researchers from one British and two Dutch universities at Loughborough in March 2007, and the calls for broader approaches to construction project management, given increasing demands for integrated teams as in partnering and alliancing, and downstream asset management as in PPPs (Kumaraswamy and Abeysekera 2007). The latter cited Black (2006) in conveying the need for managers of construction projects to be sensitive to the 'business case' of the overall project. Furthermore, they concluded by making a case for developing a body of knowledge in construction project management (CPM-BOK).

Collaborative brief development and integrated teams

Front-end project feasibility analysis and definition, as well as the technical imperatives and social dynamics in developing an appropriate (optimal would be too elusive) project brief and design brief, are worth focusing upon, given their impact on the rest of the project. The importance of the brief is reinforced by the November 2006 report to the Strategic Forum on the May 2006 Integration Workshop held by Constructing Excellence (2006), which found that the UK construction 'industry is nowhere near meeting the 50% (integrated teams and supply chains) target by end 2007'. The first of the top 10 barriers against meeting this target related to clients was identified as 'lack of clear brief, incomplete/changeable brief and lack of understanding', while the second of the top 10 success factors was concluded to be 'project brief clarity and focus on delivery value rather than lowest cost'. Other success factors for integration included 'leadership from clients and/or powerful supply team members' (top factor), appropriate team selection, early involvement of supply partners, shared objectives, shared values, openness, communication systems and continuity of work. These other success factors are also relevant to the next section of this chapter.

Focusing on the brief, it is evident that much needs to be done in involving all stakeholders early in the briefing process, given the embarrassing U-turns in some recent megaprojects in Hong Kong, for example the Central reclamation, the Lok Ma Chau spur line and the West Kowloon Cultural District project. In fact Chinyio (2007) uses stakeholder theory to argue that value can be maximised through an optimal brief if the views of the stakeholders are coordinated systematically, while research into developing collaborative briefing systems is ongoing in Hong Kong universities.

Targeting desirable project cultures

The importance of fresh approaches to 'enable team cultures' is reiterated by Fellows *et al.* (2007). Extending this concept, it is argued here that having an agreed (and 'satisficing', if not optimal) brief and the proactive generation of an appropriate project team culture can be as important as the selection of a suitable project procurement and delivery system. Kumaraswamy and Rahman (2006) mapped the multiple sources of a typical construction project team culture (e.g. from different organisations, disciplines, professions and dominant individuals) with a view to such proactive 'pushing' of the 'resultant culture' towards a desirable target, as in Figure 16.1. Kumaraswamy *et al.* (2007a) expanded on this concept and recommendation, adding further sub-culture sources in the case of longer-term PPP projects.

Relational contracting and relationally integrated value networks (RIVANs)

Relational contracting and integrated teams

Industry initiatives towards integrating long-fragmented construction project teams commenced with moves towards reintegrating specific functions such as design and construction, as well as integrating problem solving through 'partnering' and risk–reward sharing through 'alliancing'. While collaboration through partnering and alliancing have brought benefits in some countries, others are apprehensive of potential dangers in demolishing the barriers between clients' representatives and contractors (and indeed within the supply chain), which also appears contrary to traditional adversarial contracts.

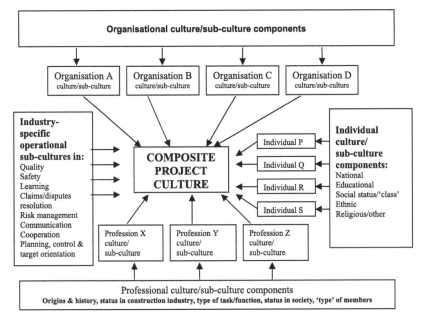

Figure 16.1 Sources of a typical construction project team culture.

However, the theoretical underpinnings for partnering- and alliancing-style collaboration have recently emerged through the advocacy of applying 'relational contracting' to construction project scenarios. A five-country survey (Kumaraswamy *et al.* 2005; Rahman *et al.* 2005) found strong support from industry in principle for the development of a 'relational contracting culture' and 'relationally integrated teams'.

Still, it is not easy to move from traditional (classical) contracts to purely relational contracts overnight. There will usually be a mix between the classical contracting and relational approaches in most scenarios. Figure 16.2 illustrates the push–pull forces that these can generate between any two, or any three, team members. As indicated, the team can be drawn to work more closely together in a more collaborative integrated mode, where the relational forces are stronger, or indeed vice versa.

Of course megaprojects involve multiple stakeholders, as well as many functions (such as financing, design and marketing). The complexity increases in PPP projects where functions extend to operation and maintenance, multi-disciplinarity in each function is greater, lifecycle considerations are more important, and project teams are expected to work together for much longer. Figure 16.3 illustrates the interacting force-fields within such an expanded multi-functional framework, with a clearly more complex supply chain.

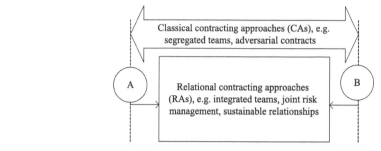

2(a) <u>Equilibrium</u> of push–pull forces between any two team members

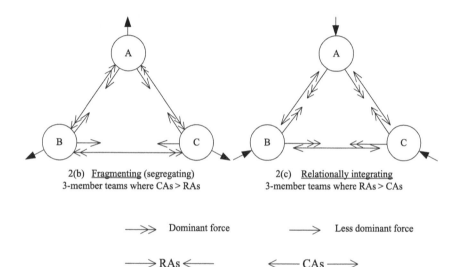

Figure 16.2 Pushing team members apart or pulling them closer together.

Relationally integrated value networks (RIVANs)

In this context, an ongoing research exercise is aiming to develop a framework for incentivising and orienting supply chains towards agreed 'best value' targets. Drawing on relevant success factors, while avoiding common pitfalls in partnering and alliancing, this exercise aims to help boost project performance, as well as long-term relationships (as in 'framework agreements' initiated by the British Airports Authority, etc.). The overall target is for more committed 'relationally integrated value networks' (Kumaraswamy *et al.* 2003) that would synergise the value streams of each network member (Kumaraswamy and Rahman 2006) more efficiently and effectively than in traditional supply chains. Figures 16.4 and 16.5 indicate the conceptualisation of a 'relationally integrated value network' and the convergence of the value streams respectively.

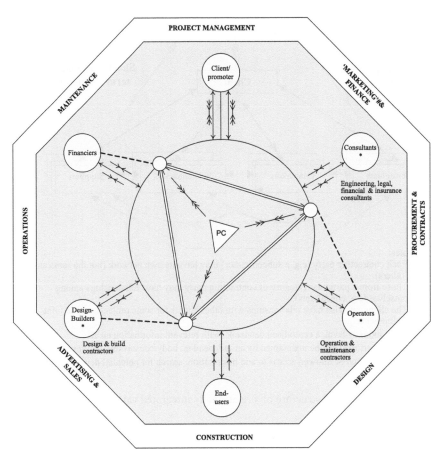

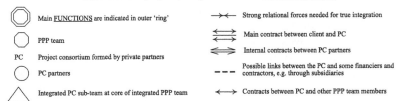

Figure 16.3 Force-fields between team members within a multi-functional PPP framework.

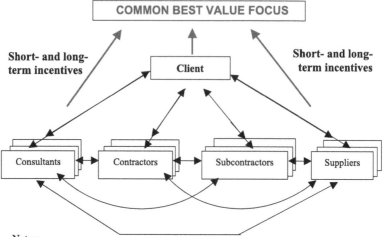

Notes:
1 Each contracting party (e.g. a subcontractor) may have its own network (not the same as that of others).
2 Those from a particular category of contracting party may have relationships among themselves (e.g. consultants).
3 The client may also have relationships with other clients for some projects, e.g. in joint developments.
4 At an industry level, a centralised databank (with relevant information on all potential partners) can be maintained by an independent body/consortium. Authorised parties may then be allowed access to add information, search for potential new partners, etc.

Figure 16.4 Proposed structure of a relationally integrated value network (RIVAN).

Management support systems

Top-down (client) systems

Large construction clients can benefit from decision aids and experiential knowledge, in designing appropriate project-specific procurement and delivery systems, for achieving synergies through relational integration as advocated above, as well as higher performance levels in general. Integrated management support systems have been proposed for large construction clients (Kumaraswamy *et al.* 2006).

Figure 16.6 shows the four primary systems that are conceptualised to aid the development of a large construction client. These include institutional, human resource and technological development, which would help in dealing with multi-project portfolios in the long term. The complexities of multiple short- and long-term performance goals point to an ICT-AI (information and communication technology – artificial intelligence)-enabled overarching management support system (MSS) to synergise optimising decision making in each system rather than to generate conflicts, for example through a downstream delivery system that is incompatible with a selected

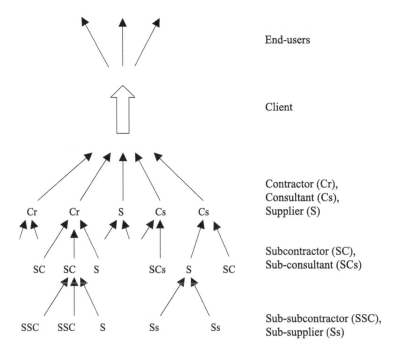

Notes:
  1 Each arrow represents the principal 'added value' contributed by each project team member.
  2 Only the principal value flow is shown. Complementary lateral flows that reinforce/add to the value of other team members are not shown here.

Figure 16.5 Conceptualisation of a project value stream.

upstream procurement strategy. This conceptualisation also incorporates provisions for longer-term human resource and technological development systems.

Procurement itself is taken to consist of five upstream strategic sub-systems, while 'delivery' incorporates the downstream managerial/operational sub-systems, shown in Figure 16.7. For example, in the procurement system, after the front-end demarcation of a megaproject into work packages (WP), the functional grouping (FG) involves decisions on whether the design, construction, project management and financing (and various components of all these functions) are handled separately by different organisations or integrated to different degrees, for example in different types of design and build, PPP or management-led systems.

Bottom-up information and knowledge management systems

While the top-down system is intended for large clients, Kumaraswamy *et al.* (2006) also describe the bottom-up SMILE-SMC information and

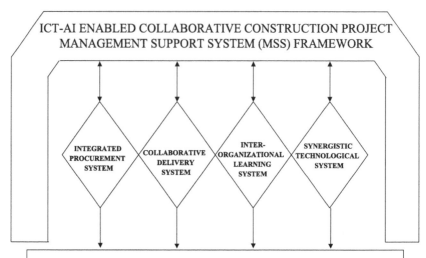

Figure 16.6 MSS model for collaborative decisions and development.

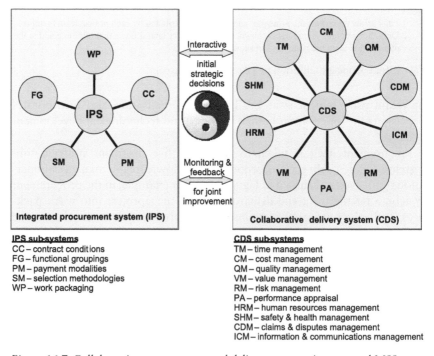

IPS sub-systems
CC – contract conditions
FG – functional groupings
PM – payment modalities
SM – selection methodologies
WP – work packaging

CDS sub-systems
TM – time management
CM – cost management
QM – quality management
VM – value management
RM – risk management
PA – performance appraisal
HRM – human resources management
SHM – safety & health management
CDM – claims & disputes management
ICM – information & communications management

Figure 16.7 Collaborative procurement and delivery systems in proposed MSS.

knowledge management system developed for small and medium contractors. An overarching project information and knowledge management platform linking the above two systems is also proposed, to mobilise and synergise groups across the project supply chains in respect of different types of work information flows and knowledge sharing, including commercial aspects. For example, linking the systems of consultants and main contractors with those of subcontractors and suppliers would accelerate the transmission of design changes to those who 'need to know', while a wider range of suppliers may be brought on-stream as well. However, it is noted that many barriers need to be crossed in information and knowledge exchange before such top-to-bottom integration can be transformed from concept to viable working system.

Technology and knowledge exchange (TKE)

The need for TKE

Since many other chapters in this book focus on significant advances in hard systems and technologies, this section focuses on how improved relationships and integrated teams, as targeted in the above sections, can lower the barriers and facilitate the exchange of technical know-how, as well as other knowledge. Synergistic technological systems were previously identified as one of the four primary systems to be developed and integrated in the MSS of a large client, as in Figures 16.6 and 16.7. However, such synergies cannot arise in the absence of very good relationships.

For example, technology transfers (TTs) have been long advocated as a way of boosting the technological capacities and construction industries of less developed countries (Ofori 1994; Simkoko 1995). However, obstacles to TTs have nullified both the good intentions and even some contractual provisions that sought to facilitate TTs (Carrillo 1995; Kumaraswamy 1995). This is not surprising, given the apparent lack of direct benefits to the suppliers of technology, and indeed the generation of future competitors. This dilemma led Kumaraswamy (1995) to conceptualise a two-way process of technology exchange (TE) that could supersede the less attractive one-way TT. TE was conceptualised on the basis of the holistic model of technology proposed by the UNESCAP-sponsored 'Technology Atlas' project of the Asia Pacific Center for the Transfer of Technology (1989). The basic model postulated four primary dimensions or components of technology as: Technoware (hardware and physical facilities), Humanware (people and their abilities), Infoware (documented facts) and Orgaware (organisational networks and managerial frameworks). This model was presented in a basic 'THIO diagram' plotted across the above four dimensions.

Kumaraswamy (1995) extended this basic THIO representation to show that some (and not necessarily all) components of a technology may be transferred at any given time or on a given project. Secondly, technology flows need not be just in one direction: for example, one (or two) components may be transferred from party A to party B and vice versa. This reinforces the rationale of joint ventures (JVs) where the strengths of one party compensate for the weaknesses of the other.

The overall integrated JV technology profile between two partners A and B can possibly expand to Y, as shown in Figure 16.8. However, inappropriate structuring of the JV functions, unrealistic resource demands and/or risk–reward mechanisms could diminish the integrated technology profile to X, that is, to levels even below A's or B's in respect of particular technology components. A previous comparison of 'integrated' and 'non-integrated' JVs in Singapore found that structural integration of the JV is crucial to JV success and continuity (Sridharan 1994). Here, integration referred to the actual working together on specific work packages rather than compartmentalising and allocating each package to different partners, hence implying 'structural' integration of the organisational and working arrangements.

The current chapter argues that 'relational integration' of the JV teams is as critical as structural integration, since it can create a culture conducive to TE. This can start by first nudging organisations into a 'knowledge sharing' mode to capture, develop and spread knowledge within the organisation in the first instance (Carrillo *et al.* 2003). Pushing them further to share

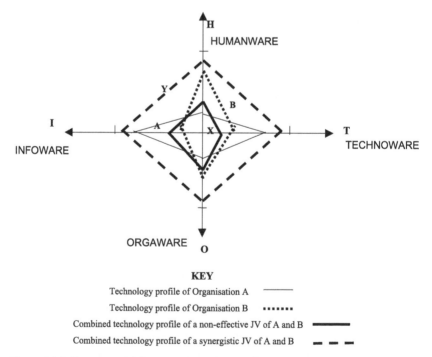

Figure 16.8 Two potential scenarios of overall (combined) JV capacities and competencies.

knowledge with JV partners should then be easier than before, that is, to migrate from an intra-organisational to an inter-organisational learning mode. Given the growing popularity of knowledge management, it is easier to link the broader visualisation of 'technology' to the even broader conceptualisation of 'knowledge' that highlights the deeper and longer-term experiential facets. This should reinforce such exchange exercises further – hence the extension to TKE in this chapter.

It is worth noting that 68 per cent of 150 respondents in a 2003 survey in Hong Kong perceived TT to be useful for their organisation and 74 per cent felt it is useful for the industry, whereas only 42 per cent felt it was taking place effectively in JVs at present. For this to improve, the many barriers to TT must be surmounted. The top five barriers to TTs had been identified in a previous survey (Kumaraswamy and Shrestha, 2002) to be: organisational culture, lack of time, lack of capacity, individual attitudes and lack of clear policy.

Another Hong Kong-based study (Choi 2005) extended the TE concept to TKE as deemed feasible above and carried out an interesting survey to test its acceptability to practitioners. The survey yielded 65 useful responses, with 19 per cent being from public clients, 22 per cent from consultants, 20 per cent from contractors and 39 per cent from sub-contractors. This study also extended the application of TE/TKE from JVs to construction supply chains. The perceived value of TKE was clearly confirmed in this survey, with specific views derived in terms of each of the various components (T, H, I and O) and also the more detailed contributors/factors relating to each. For example, when taking all responses together, it was seen that 54 per cent, 57 per cent, 22 per cent and 13 per cent broadly agreed that TKE takes place now in terms of T, H, I and O respectively. This indicates less 'exchange' of information and less interaction/integration of the partner organisations. Secondly, information on perceived barriers and drivers to TKE in particular was also elicited, for example 84 per cent of respondents believed that the 'cultural gap' between organisations must be bridged.

JVs provide a good starting point for building integrated teams, and are more suited for TE, given the presumably 'equal' status of the base relationships, as compared to subcontracts, for example. JVs are also increasingly common in construction industries, given the increasing complexity and magnitude of infrastructure projects that need diverse resource inputs from multi-disciplinary teams.

Figure 16.9 visualises the need for closer integration along each of the H, I and O dimensions in order to achieve meaningful TKE. It highlights (on top) the need for relational integration along the H dimension, and incorporates the need for structural integration along the O dimension. ICT-aided collaborative systems (in the I dimension) will further facilitate both relational and structural integration, as well as collaboration, as in Figures 16.6 and 16.7.

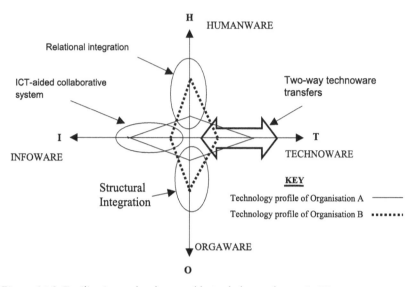

Figure 16.9 Facilitating technology and knowledge exchange in JVs.

Kumaraswamy *et al.* (2004) expanded on how 1) relational integration and 2) ICT-aided collaborative systems would complement 3) any structural integration at organisational level, as indicated above. All three may be hypothesised as interdependent prerequisites to ensuring effective transfers of technoware. However, a comprehensive evaluation model is needed for assessing the mutual transfers of such soft and hard factors, in the envisaged TKE.

Cooperation, competition and coopetition

The above sections and sub-sections argue for what some may consider to be altruistic, if not idealistic, relationships, cooperation and sharing arrangements. In this context, it may also be noted that the CIB W107 developed definition of TT is: 'TT has been said to have taken place where the receiver can use and adapt technology components on a self-reliant and sustainable basis in the construction/production processes', that 'the receiver has gained capabilities to adapt, maintain and even generate "technology components" more or less based on the originally received ones' and that 'in the end the receiver might be a competitor of the supplier' (Ofori 2001).

Still, competition is not incompatible with cooperation, as evidenced by emerging literature on coopetition. Accepted drivers towards competitive advantage, as popularised by Porter (1985), have been challenged by Elmualim *et al.* (2006), who identified 'counter-discourses', including one stemming from moves towards corporate social responsibility and another lamenting 'hollowed-out firms' in the UK that failed to invest in their human capital in order to be structurally flexible to cope with demand fluctuations for competitive advantage.

Meanwhile, coopetition had been conceptualised by Brandenburger and Nalebuff (1996) as 'co-operation to compete better'. This conceptualisation was applied by Mahesh and Kumaraswamy (2006) to examine whether JVs in Indian infrastructure, such as in airports, arose from a coopetitive strategy or 'coalitions of necessity'. In the latter context, Anvuur and Kumaraswamy (2008) differentiated short-term project-oriented (and thereby specific-objective-driven) collaboration from general cooperation, and showed that better collaboration could be achieved through genuine cooperation, while Anvuur and Kumaraswamy (2007) provided a conceptual model of partnering and alliancing that can help project managers focus on teamworking processes that lead to high cooperation and performance levels.

Other interpretations of coopetition include: 'co-opetition involves co-operating to create a bigger business pie, while competing to divide it up' and 'there may be more co-operation in some geographical areas, product domains or value chain functions, while more competition may be involved in others' (Luo 2004). Levy *et al.* (2003) noted how coopetition or simultaneous cooperation and competition, may aid competitiveness by knowledge sharing, but also how exchanged knowledge may be used for competition, therefore calling for better skills and IS/IT systems in managing inter-organisational knowledge flows.

Given the various general interpretations and emphases as above, the need is noted for specific research on coopetition in the context of construction project management, for example examining how relationally integrated value chains can remain competitive *vis-à-vis* the rest of the industry (i.e. avoiding complacency, inefficiencies and even potential collusion/corruption arising from cosy coalitions) while achieving cooperative synergies.

Conclusions

The foregoing sections provide insights into a range of related soft system issues and initiatives, with a view to better utilisation of the rapid advances in hard systems that can be deployed in providing infrastructure for the built environment. While mostly presented from organisational and project perspectives, these must also be considered in the broader and longer-term national, regional and indeed global contexts.

There will of course be differences in approaches and applications. For example, Kumaraswamy *et al.* (2007b) presented an industry-level perspective for 'revaluing construction' from the viewpoint of a range of developing countries, within the overall CIB priority theme of 're-valuing construction' in the 'organisation and management of construction' context. However, common agendas do emerge, as in the 'Conclusions and Recommendations' by Ofori and Milford (2007) following the May 2007 CIB Congress. These recommendations include:

As a community of researchers, we recognized that we should:

- Develop greater understanding of the relationships between construction and development. . . .
- Adopt a broader perspective to our research, and adopt multidisciplinary approach, embracing also social development.
- Move forward our work on sustainable development. . . .
- Establish a common platform of meanings and implications of the same term or expression. . . .
- Engage industry and endeavour to make our work relevant to the needs of industry.

In this broader context, it is felt useful to develop and present a preliminary conceptual map in Figure 16.10, both to capture and compare some of the issues discussed in this chapter and to position them on a broader canvas, for future investigation and development.

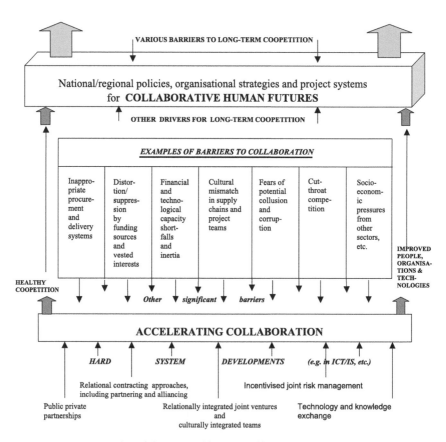

Figure 16.10 Examples of drivers and barriers affecting collaboration.

References

Anvuur, A. M. and Kumaraswamy, M. M. (2007). 'A Conceptual Model of Partnering and Alliancing', *Journal of Construction Engineering and Management*, ASCE, 133(3), March, pp. 225–234.

Anvuur, A. M. and Kumaraswamy, M. M. (2008). 'Better Collaboration through Cooperation', in H. J. Smyth and S. D. Pryke (eds), *Collaborative Relationships in Construction: Developing Frameworks and Networks*, pp. 107–128. Oxford: Wiley-Blackwell.

Asia Pacific Center for the Transfer of Technology (1989). *A Framework for Technology-Based Development Planning*, Vol. 2: *Technology Content Assessment*. Bangkok, Thailand: United Nations Economic and Social Commission for Asia and the Pacific (UNESCAP).

Black, R. J. (2006). 'Project Managing the Business Cases as well as Construction', *Proceedings of the One Day Seminar on Recent Developments in Project Management in Hong Kong*, 12 May 2006, Hong Kong, Centre for Infrastructure and Construction Industry Development, University of Hong Kong (CICID HKU), pp. 3.1–3.4.

Brandenburger, A. M. and Nalebuff, B. J. (1996). *Co-opetition*. New York: Doubleday.

Bresnen, M., Swan, J. and Goussevskaia, A. (2005). Special issue on 'Managing Projects as Complex Social Settings', *Building Research and Information*, 33(6), Routledge, UK.

Carrillo, P. (1995). 'Technology Transfer on International Joint Venture Projects', *Proceedings of First International Conference on Construction Project Management*, pp. 327–334, Nanyang Technological University, Singapore, January 1995.

Carrillo, P., Robinson, H. and Hartman, F. (2003). 'Knowledge Management Strategies: Learning from Other Sectors', *Proceedings of the 2003 ASCE Construction Research Congress*, Hawaii, USA. CD Rom.

Chinyio, E. (2007). 'A Shared Project Brief', in M. Sexton, K. Kahkonen and S. Lu (eds), *Revaluing Construction: A W065 'Organisation and Management of Construction' Perspective*, CIB Publication 313, May, pp. 12–26. Rotterdam, Netherlands: CIB.

Choi, Y. P. (2005). 'Technology and Knowledge Exchange in Construction Project Supply Chains', M.Sc. project dissertation, Department of Civil Engineering, University of Hong Kong, Hong Kong.

Constructing Excellence (2006). *Towards More Effective Integration in UK Construction*, Report following the Integration Workshop held on 10 May 2006 at the DTI Conference Centre, London, http://www.constructingexcellence.org.uk/pdf/ce_integration_workshop_report_oct_2006.pdf

Elmualim, A. A., Green, S. D., Larsen, G. and Kao, C. C. (2006). 'The Discourse of Construction Competitiveness: Material Consequences and Localised Resistance', in M. Dulaimi (ed.), *Joint International Conference on Construction Culture, Innovation and Management*, pp. 446–456. Dubai: CIB.

Erdogan, B., Aumba, C. J., Bouchlaghem, D. and Nielsen, Y. (2006). 'Organisational Change Management for Collaborative Construction Environments', in H. Rivard, E. Miresco and H. Melhem (eds), *Joint International Conference on Computing and Decision Making in Civil and Building Engineering*, pp. 1393–1402, Montreal, Canada: CIB.

Fellows, R., Grisham, T. and Tjihuis, W. (2007). 'Enabling Project Team Cultures', in M. Sexton, K. Kahkonen and S. Lu (eds), *Revaluing Construction: A W065 'Organisation and Management of Construction' Perspective*, CIB Publication 313, May, pp. 27–44. Rotterdam, Netherlands: CIB.

Flanagan, R. (2007). 'The Drivers and Issues Shaping the Construction Sector', in Y. M. Xie and I. Patnaikuni (eds), *4th International Structural Engineering and Construction Conference (ISEC-4) on 'Innovations in Structural Engineering'*, Melbourne, Australia, 26–28 September, Vol. 2, pp. 3–11. London: Taylor & Francis.

Kumaraswamy, M. M. (1995). 'Synergy through Technology Exchange', *1995 International Congress of Engineering Deans and Industry Leaders, Melbourne, Australia*, Proceedings, pp. 563–567, July.

Kumaraswamy, M. M. and Shrestha, G. B. (2002). 'Targeting "Technology Exchange" for Faster Organisational and Industry Development', *Building Research and Information*, 30(3), pp. 183–195.

Kumaraswamy, M. M. and Rahman, M. M. (2006). 'Applying Teamworking Models to Projects', in S. Pryke and H. Smyth (eds), *The Management of Complex Projects: A Relationship Approach*, pp. 164–186. Oxford: Blackwell.

Kumaraswamy, M. and Abeysekera, V. (2007). 'Revisiting Terminology in Construction Project Management', in Y. M. Xie and I. Patnaikuni (eds), *4th International Structural Engineering and Construction Conference (ISEC-4) on 'Innovations in Structural Engineering'*, Melbourne, Australia, 26–28 September, Vol. 2, pp. 1459–1465. London: Taylor & Francis.

Kumaraswamy, M. M., Rahman, M. M., Palaneeswaran, E., Ng, S. T. and Ugwu, O. O. (2003). 'Relationally Integrated Value Networks', in C. Anumba (ed.), *Proceedings of 2nd International Conference on Innovation in Architecture, Engineering and Construction*, pp. 607–616, Loughborough, UK, June 2003.

Kumaraswamy, M. M., van Egmond, E. L. C., Rahman, M. M. and Ugwu, J. (2004). 'Technology Exchange through Relationally Integrated Joint Venture Teams', *CIB W92 International Symposium on Procurement Systems*, pp. 326–334, Chennai, India, January.

Kumaraswamy, M. M., Ling, F. Y. Y., Rahman, M. M. and Phng, S. T. (2005). 'Constructing Relationally Integrated Teams', *Journal of Construction Engineering and Management*, ASCE, 131(10), pp. 1076–1086.

Kumaraswamy, M. M., Palaneeswaran, E., Rahman, M. M., Ugwu, O. O. and Ng, S. T. (2006). 'Synergising R&D initiatives for e-enhancing management support systems', *Automation in Construction*, Special issue on 'Knowledge Enabled Information Systems Applications in Construction', 15(6), September, pp. 681–692.

Kumaraswamy, M., Rahman, M., Palaneeswaran, E., Ugwu, O., Anvuur, A. and Yogeswaran, K. (2007a). Multi-disciplinary and Multi-functional Teams in PPP Procurement and Delivery', in K. London, G. Thayarapan and J. Chen (eds), *CIB W092 Procurement Systems Conference on 'Building across Borders'*, pp. 250–262, University of Newcastle, Newcastle, Australia, 23–26 September.

Kumaraswamy, M., Lizarralde, G., Ofori, G., Styles, P. and Suraji, A. (2007b). 'Industry-Level Perspective of Revaluing Construction: Focus on Developing Countries', in M. Sexton, K. Kahkonen and S. Lu (eds), *Revaluing Construction: A W065 'Organisation and Management of Construction' Perspective*, CIB Publication 313, May, pp. 88–103. Rotterdam, Netherlands: CIB.

Levy, M., Loebbecke, C. and Powell, P. (2003). 'SMEs, Co-opetition and Knowledge Sharing: The Role of Information Systems', *European Journal of Information Systems*, 12, pp. 3–17.

Luo, Y. (2004). *Co-opetition in International Business*. Copenhagen, Denmark: Copenhagen Business School.

Mahesh, G. and Kumaraswamy, M. M. (2006). 'Joint Ventures in Indian Infrastructure Industry: Co-opetitive Strategy or Coalitions of Necessity?', *2nd International Conference on Multi-national Joint Ventures for Construction Works*, pp. 159–165, Hanoi, Vietnam, 28–29 September. Hanoi: Construction Publishing House, Vietnam.

Morris, P. W. G. (2006). 'How Do We Learn to Manage Projects Better?', in S. Pryke and H. Smyth (eds), *The Management of Complex Projects: A Relationship Approach*, pp. 58–77. Oxford: Blackwell.

Ofori, G. (1994). 'Construction Industry Development: Role of Technology Transfer', *Construction Management and Economics*, 12, pp. 379–392.

Ofori, G. (2001). 'CIB TG29: Construction in Developing Countries', Progress Report 1997–2000, Singapore, July.

Ofori, G. and Milford, R. (2007). 'Conclusions and Recommendations', CIB World Building Congress 2007, Cape Town, South Africa, CIB News Article, CIB Management, Rotterdam, September.

Porter, M.E. (1985), *Competitive Advantage*. New York: Free Press.

Rahman, M. M., Kumaraswamy, M. M., Karim, K., Ang, G. and Dulaimi, M. (2005). 'Cross-country Perspectives on Integrating Construction Project Teams', 6th Construction Specialty Conference, Canadian Society of Civil Engineers, Toronto, Canada, 2–4 June. CD Rom.

Simkoko, E. E. (1995). *Analysis of Factors Impacting Technology Transfers in Construction Projects*. Stockholm: Swedish Council for Building Research.

Sridharan, G. (1994). 'Managing Technology Transfer in Construction Joint Ventures', *American Association of Cost Engineers Transactions*, Vol. 1994, pp. 6.1–6.4. Morgantown, WV: AACE.

17 The construction game

Ghassan Aouad, Song Wu and Angela Lee

Abstract

Technologies for video games have advanced dramatically over the past decade, revolutionising the motion picture, game and multimedia industries. Similar technologies, such as virtual reality and digital modelling, have been used in the construction domain; examples can be found in urban planning and city modelling. The nD game project, by the University of Salford, developed a computer game to inspire younger generations into the construction industry. With modern games technology becoming more and more sophisticated, and the vast amount of technologies that have been developed during the game development, this chapter explores the opportunities to apply gaming technology in the construction arena. Technologies such as game artificial intelligence (AI) and behavioural modelling for construction simulation, massively multiplayer online game (MMOG) for the virtual construction site, and interactive games for construction training are discussed.

Introduction

The idea of a construction game is not new; it has been used for some time to help students develop construction project management skills. Au and Parti (1969) first suggested that computerised heuristic games could be used for the education of engineers and planners engaged in the construction industry. A scheduling game was developed by Scott and Cullingford (1973) in the 1970s. Halpin (1976) presented a model for a project gaming system that trains students to plan, monitor and control hypothetical projects. Harris and Evans (1977) developed a road construction simulation game for site managers, focusing on the planning and control of linear construction projects. Herbsman (1986) explained the use of civil engineering project management games in the US, where players were required to participate in the design and execution phase of live projects. Rounds *et al.* (1986) described a construction game that simulates the progress and project reporting structure of an industrial construction project. Dudziak and

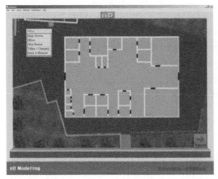 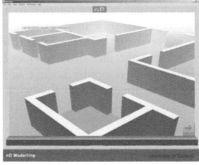

Figure 17.1 nD game.

Hendrickson (1988) developed a game for contract negotiations. Vehosky and Egbers (1991) explained the development and use of a simulation game for the management of the design of a construction project. AbouRizk (1993) and AbouRizk and Sawhney (1994) described the development and deployment of a construction bidding game that provides undergraduate civil engineering students with a thorough understanding of the components and methods of bidding. Scott *et al.* (2004) described the use and evaluation of a web-based simulation game to teach planning and control to undergraduate construction engineering students.

Most recently, the nD game (Aouad *et al.* 2007) developed by Salford University was designed as an educational interactive multimedia learning aid directed at 11- to 14-year-olds (Figure 17.1). It sets school pupils, in groups, to enact differing roles in the construction industry, with the task of designing their own school. The objective of the game is to highlight the complexity of the design process, and to encourage the next generation of construction professionals. The project was funded by the Engineering and Physical Sciences Research Council (EPSRC) under a Public Participation (P3A) award.

Most of the systems mentioned above, however, either are manual or have been developed with limited use of advanced game technology. As the development of video games is advancing rapidly, many modern video games draw from a number of disciplines, including media studies, education and learning theory, human–computer interaction (HCI), psychology, sociology, anthropology and so on. Technologies such as game AI or behaviour modelling and MMOG have been developed. This chapter aims to introduce the technologies and explore their potential applications in the construction industry.

Game artificial intelligence and behaviour modelling

Game artificial intelligence refers to techniques used in computer and video games to produce the illusion of intelligence in the behaviour of non-player characters (NPCs). The process of producing the intelligence in the behaviour is often referred to as behaviour modelling. The techniques used typically draw upon existing methods from the academic field of AI. However, the term 'game AI' is often used to refer to a broad set of algorithms that also include techniques from control theory, robotics, computer graphics and computer science in general (Wikipedia 2007a). Typical AI techniques are used in the following areas.

Navigation

NPCs have to move through the environment in an intelligent way, that is, not getting stuck by trees in their way, taking a possibly short route to the destination, and so forth. The so-called A* algorithm is the most common basic technique for path finding for an NPC. The algorithm requires a definition of waypoints and their connections for a specific environment. Given a starting point and a destination, the A* algorithm tries to find the shortest path along the waypoint connections (Nareyek, 2004).

Team and formation movement can also be incorporated in a similar way. For example, a general movement vector per team is introduced, which is combined into each team member's vector set, as well as a vector for each member that points toward the preferred position within the team. In many games and movies, flocking birds or fishes in the background are also realised by these techniques.

Similar techniques can be used in building fire safety design for a fire escape route. Techniques used in team and formation movement can also be seen in the crowd simulation for an emergency event or to simulate crowd dynamics in an urban area.

Decision making

The most common representations for decision-making processes in game characters are finite-state machines (FSMs). Character behaviours are modelled as a finite set of states, with transitions between them in the form of a directed graph. The character resides in only one by the conditions and actions occurring in the game. There are a number of extensions to FSMs which make it possible to develop more complex behaviours, including stack-based FSMs, fuzzy state machines, hierarchical FSMs, and probabilistic FSMs (Nareyek 2004). However, current FSM techniques do not scale well, and their propositional representation can make behaviours difficult to represent (Diller *et al.* 2004).

Rule-based approaches to behaviour have also been implemented in games. Typically, in these systems a set of if-then expressions is used to encode simple condition-action pairs. Recently, a few games have developed character behaviours using goal-directed reasoning techniques. In this technique game characters have a set of goals, and actions are chosen in order to best satisfy the most relevant goal or goals. Finally, there are a small number of games which have used neural networks or genetic algorithms to specify behaviours.

Sensory mechanisms

The sensory mechanisms by which game characters 'see' and in some cases 'hear' the world range in complexity from the extremely simple to the surprisingly sophisticated and complex. Compared to the rich and compelling virtual worlds provided to human players, the world representations designed for synthetic characters are quite impoverished. Environments are typically reduced to simplified representation used only for navigation. Object detection occurs by performing collision tests radiating out from the NPC's location. Sensory routines often perceive only entities and events that can influence the NPC's reactions (e.g. Halo) – typically human players and the actions they produce. Routines are run periodically in order to gather relevant events. Vision is often represented by a two-dimensional view cone, and a line-of-sight operator is used to detect whether an object is visible to the NPC. Unfortunately, line-of-site detection does not take into account real-world cues such as motion detection, depth perception, and coloration or texture differences. Hearing is often implemented using even simpler models. For example, Halo models the distance that an NPC can hear a sound as a function of the volume of the sound scaled by a 'hearing sensitivity' parameter. Smell is sometimes implemented using the same mechanisms as for sound (Diller *et al.* 2004).

Learning

A number of AI learning techniques have been employed in computer games, including decision trees, neural networks and genetic algorithms. Most of these techniques are used during the game-development process, and few games actually incorporate the learning mechanisms for fear that the game will learn poorly and provide a less than desirable game-play experience (Nareyek 2004). There are notable exceptions, however. Black & White has learning as a central component in the game. In Black & White, a player controls a Creature that learns from the player's actions and any positive or negative feedback provided by the player. The Creature, based on the belief–desire–intention agent architecture, learns using a variety of learning algorithms, including decision trees and neural networks (GameAI.com 2007).

Relatively few games have used genetic algorithms. A notable exception is Cloak, Dagger, and DNA, which used genetic algorithms to evolve opponent strategy. Between battles, the player can allow the opponent's DNA strands to compete and evolve through a series of tournaments (GameAI.com 2007).

Although all these techniques developed in the game industry aimed to make games appear to be intelligent rather than actually be intelligent (Diller *et al.* 2004), the commercial games did provide extensive examples of how simple AI behaviour can be used effectively in virtual worlds, for the developer of construction simulation or training applications may be able to take advantage of the techniques to make the simulation or training application more intelligent, or close to the reality. For example, intelligent avatars can be developed to assess the accessibility and security of a new building design. The behaviour of the avatars can be modelled by using the techniques introduced earlier.

There is a key similarity between the development of intelligent behaviour in games and training or simulation systems. That is the need to facilitate cooperation between the content experts who describe and specify the desired behaviours and the programmers who implement them. However, the behaviours will be pre-defined and hard-coded, and the avatars will only behave accordingly. The approach will not be successful if the behaviours are difficult to model, such as human feelings and human experiences. There is another form of game virtual environment which is evolved and controlled by users. It is called massively multiplayer online game.

Massively multiplayer online game

MMOG is a computer game which is capable of supporting hundreds or thousands of players simultaneously. It is played on the Internet and features at least one persistent world. A persistent world is a virtual world which is always available and events happen continually. One of the most well-known MMOG games is Second Life.

Second Life is an Internet-based virtual world launched in 2003, developed by Linden Research, Inc. (commonly referred to as Linden Lab). A downloadable client program called the Second Life Viewer enables its users, called 'residents', to interact with each other through motional avatars, providing an advanced level of social network service combined with general aspects of a metaverse. Residents can explore, meet other residents, socialise, participate in individual and group activities, and create and trade items (virtual property) and services with one another. Second Life has its own virtual currency, the Linden dollar, which is exchangeable for US dollars in a marketplace consisting of residents, Linden Lab and real-life companies. In all, more than 8.9 million accounts have been registered in the virtual world (Wikipedia 2007b).

Real-life architectural firms are also starting to explore the use of Second Life to enhance their real-world business. It is especially appealing for architects looking for feedback on buildings long before construction begins. For example, Crescendo Design showed how it uses its virtual land in Second Life to build virtual models of its design concepts and redesigns the virtual site so that it is similar to the actual site. It then lets clients access this virtual model in Second Life and spend as much time in it as they want, get a sense of the design and provide their feedback. It conducts meetings in virtual 'real time' where both the architect and the client meet in the form of their respective avatars at the virtual site and tour it together. It also uses its virtual studio as an educational model which describes the value of several green design features (Khemlani 2007).

In addition to being an alternative design and presentation medium for architects, Second Life can also emerge as a useful educational tool. In San Francisco, an architecture professor uses Second Life to teach collaboration, and students develop their design in the virtual world (Wong 2007). At the Montana State University School of Architecture, students are using Second Life's group creation platform, and they can work in the same interface to manipulate geometric shapes and link them to make a variety of structures. It was found that it is a good tool for showing students how building parts fit together (Kieran 2007).

Within Second Life, the Arch is dedicated to exploring the convergence of virtual architecture with real-world architectural practice. It also describes other examples of classic works being recreated in Second Life, such as Palladio's Villa Venete. If all the architectural masterpieces are re-created online, they will be experienced in a far more interactive manner than with photographs or videos alone (Khemlani 2007).

However, there are still limitations in the current form of Second Life. One of the biggest problems is that there is no easy way to transfer building models created in real-world practice into the virtual world of Second Life. The building in Second Life has to be created using its internal modelling system. This could be a huge barrier for the construction industry to entering this powerful and exciting virtual world. There are attempts by researchers to develop tools to bridge this gap, but they are still at a very early stage. It would be excellent if the vendors of modelling tools used in AEC could work with the creators of virtual worlds like Second Life to ensure that building models created in professional practice could be seamlessly published in the virtual world, complete with all their details and textures. When that happens, Second Life can really take off in the construction industry (Khemlani 2007).

Apart from Second Life, other forms of MMOGs have been designed to accurately simulate certain aspects of the real world. They tend to be very specific to industries or activities of very large risk and huge potential loss, such as military simulation, aeroplanes and so on. The air traffic simulation is one example, with networks such as VATSIM and IVAO striving to provide rigorously authentic flight-simulation environments to players in both pilot and air traffic controller roles (Vatsim.net 2007). The United States Army is developing a massively multiplayer training simulation called 'asymmetric warfare environment' (AWE) to train soldiers for urban warfare (Gonzalez 2004).

Many characteristics in MMOG have been identified to be useful for broader application of game-based technologies. At the same time, there are a number of limitations in MMOGs, such as their support of game authorability, persistent in-game effects, and participation of a large number of simultaneous players, which are not universally acknowledged (Diller *et al.* 2004). While MMOG authorability, persistence and massiveness are potentially attractive for training or other simulation applications, developers must be aware that these characteristics are currently limited and that the commercial game industry is just starting to push these limits (Diller *et al.* 2004).

Although the development of the MMOG simulation in the construction industry is still in its infancy, there is great potential for MMOG to be used for simulating the operation of a construction site, providing training for health and safety on-site, and simulating emergency events for urban planning purposes. In architectural design, MMOG can be used for validating a new building design for its security and accessibility requirement. For construction management, MMOG can be used to simulate the design and construction process.

Summary

The computer game industry has invested heavily in the development of sophisticated game characters and achieved impressive advances in both technology and its user base. The simulation and training community has long sought to leverage game technology, particularly on military simulation applications. In the construction industry, the gaming concept for training and education has been long established, and interest in game-based technologies and their actual application is increasing. Should construction researchers jump on to the gaming bandwagon?

The authors' answer is yes. Commercial game developers share the same goals for the simulation of human behaviours as simulation or training application developers, such as creating an immersive simulated world. Game developers are increasingly relying on game AI to distinguish their game from those of competitors. At the same time, game developers have become increasingly concerned with producing realistic and robust behaviours (Diller *et al.* 2004). These needs and intentions for the development of behaviours in synthetic entities are shared with the simulation or training community.

One direct benefit is that construction simulation or training applications may be able to take advantage of simple behaviour models and easy-to-use graphical authoring tools, and still produce realistic enough behaviour. If complex models are needed, some game tools for behaviour authoring might be extensible or compatible with more complex, psychologically valid models.

This chapter sought to start a discursive argument as to the need for and benefits of introducing gaming technology into construction. It described the scope and implications of two leading game technologies, namely game

AI and MMOG. Although game AI or behaviour modelling will make construction simulation more realistic and robust, human behaviour is complex, and current technology is still at its very early stage. Only simple and basic behaviour can be accurately modelled. Another form of game environment, MMOG, can compensate for the weakness of game AI (behaviour modelling) by putting more focus on the user (player) than synthetic entities in the virtual world. The avatar in the virtual world is controlled by a human; therefore, no behaviour needs to be modelled. Second Life has demonstrated great potential for design collaboration and engaging users and gauging feedback from real users through the virtual world.

Much development work and evaluation need to take place before it can be conclusively decided if gaming technology can aid construction.

References

AbouRizk, S. M. (1993). 'Stochastic Simulation of Construction Bidding and Project Management', *Microcomputers in Civil Engineering*, 8, pp. 343–353.
AbouRizk, S. M. and Sawhney, A. (1994). 'Simulation and Gaming in Construction Engineering Education', ASEE/dE2/dEI Conference, Edmonton, Alberta, June.
Aouad, G., Lee, A. and Wu, S. (2007). *Constructing the Future: nD Modelling*. London: Taylor & Francis.
Au, T. and Parti, E. (1969). 'Building Construction Game: General Description', *Journal of the Construction Division*, 95(CO1), pp. 1–9.
Diller, D., Ferguson, W., Leung, A., Benyo, B. and Foley, D. (2004). 'Behaviour Modelling in Commercial Games', Conference on Behaviour Representation in Modelling and Simulation (BRIMS).
Dudziak, W. and Hendrickson, C. (1988). 'Simulation Game for Contract Negotiations', *Journal of Management in Engineering*, 4(2), pp. 113–121.
GameAI.com (2007). 'Games Making Interesting Use of Artificial Intelligence Techniques', http://www.gameai.com (accessed 22 August 2007).
Gonzalez, L. (2004). 'Spot On: The US Army's There-Based Simulation', http://uk.gamespot.com/news/2004/04/21/news_6093860.html (accessed 22 August 2007).
Halpin, D. W. (1976). 'CONSTRUCTO: An Interactive Gaming Environment', *Journal of the Construction Division*, 102(CO1), pp. 145–156.
Harris, F. C. and Evans, J. B. (1977). 'Road Construction : Simulation Game for Site Managers', *Journal of the Construction Division*, 103(CO3), pp. 405–414.
Herbsman, Z. (1986). 'Project Management Training Using Microcomputers', *Journal of Management in Engineering*, 2(3), pp. 165–176.
Khemlani, L. (2007). 'Exploring Second Life and Its Potential in Real Life AEC', http://www.aecbytes.com/buildingthefuture/2007/SecondLife.html (accessed 22 August 2007).
Kieran, C. (2007). 'Second Life and Google Earth Are Transforming the Idea of Architectural Collaboration', http://archrecord.construction.com/features/digital/archives/0701dignews-2.asp (accessed 22 August 2007).
Nareyek, A. (2004). 'Artificial Intelligence in Computer Games: State of the Art and Future Directions', *ACM Queue*, 1(10), pp. 58–65.

Rounds, J. L., Hendrick, D. and Higgins, S. (1986). 'Project Management Simulation Training Game', *Journal of Management in Engineering*, 2(4), pp. 272–279.

Scott, D. and Cullingford, G. (1973). 'Scheduling Games for Construction Industry Training', *Journal of the Construction Division*, 99(CO6), pp. 81–92.

Scott, D., Mawdesley, M. and Al-Jibouri, S. (2004). 'The Use and Efficacy of a Simulation Model for Teaching Project Control in Construction', First International Conference on World of Construction Project Management, Toronto, May.

Vatsim.net (2007). http://www.vatsim.net (accessed 22 August 2007).

Vehosky, D. and Egbers, J. H. (1991). 'Civil Engineering Project Management Game: Teaching with Simulation', *Journal of Professional Issues in Engineering Education and Practice*, 117(3), pp. 203–213.

Wikipedia (2007a). 'Game Artificial Intelligence', http://en.wikipedia.org/wiki/Game_artificial_intelligence (accessed 22 August 2007).

Wikipedia (2007b). 'Massively Multiplayer Online Game', http://en.wikipedia.org/wiki/Massively_multiplayer_online_game (accessed 22 August 2007).

Wong, K. (2007). 'City Turns New Leaf in Second Life', http://aec.cadalyst.com/aec/article/articleDetail.jsp?id=451110 (accessed 22 August 2007).

18 A group support system for collaborative working in a value management workshop environment

Geoffrey Qiping Shen, Shichao Fan and John Kelly

Abstract

A group support system (GSS) is an interactive computer-based information system which combines the capabilities of communication technologies, database technologies, computer technologies and decision technologies to support the identification, analysis, formulation, evaluation and solution of semi-structured or unstructured problems by a group in a user-friendly computing environment. This chapter introduces a research that has been conducted to explore the potential of using GSS in value management (VM) workshops and to investigate the effect of the application. It begins with an introduction to the problems of implementing VM in the construction industry in Hong Kong. It then proceeds to an illustration of the features of the GSS prototype, named Interactive Value Management System (IVMS). Two validation studies designed to validate the support of IVMS to collaborative working are described and the results are discussed. Findings from this research indicate that IVMS is an effective tool to overcome the major problems and difficulties of collaborative working by multiple stakeholders in VM workshops.

Keywords

Collaborative working, group support systems, interactive value management system, workshops, value management, construction.

1 Introduction

Value management is a function-oriented, systematic team approach to providing value in a product, system or service (SAVE International 1998). The process uses structured, team-oriented exercises that make explicit and appraise existing or generated solutions to a problem, by reference to the value requirement of the client (Male *et al.* 1998). The implementation of VM in a construction project is normally in the form of one or more workshops, which are attended by the major stakeholders, facilitated by a value specialist, and follow a 'systematic job plan'.

As a result of technological development, uncertain economic conditions, social pressures and fierce competition, construction industry clients are placing increasing demands upon the industry in terms of the project quality, costs of delivery, time from inception to occupation and, above all, value for money of projects. As a useful tool that can help the industry meet these challenges, value management has been widely used in many developed countries for several decades. There has been a surge of interest in VM in the construction industry in Hong Kong, especially since the Asian financial crisis in 1997 (Shen and Kwok 1999). A large number of government departments and some private enterprises in Hong Kong have applied VM to ensure value for money for their projects. This is mainly because of the technical circular issued by the Works Bureau (1998), which has been subsequently revised by the Environment, Transport and Works Bureau (ETWB 2002), which recommends VM studies be carried out for a wide range of major projects in public works. The Construction Industry Review Committee (2001) also recommends that 'VM should be used more widely in local construction, because it can help clients and project teams focus on the objectives and needs of the project and all stakeholders, both long-term and short-term'.

As suggested by Nunamaker *et al.* (1997), many things can go wrong with teamwork and, as a result, the group decision-making process can be very difficult and unproductive. A recent survey suggested that VM users in Hong Kong encounter the problems of lack of active participation and insufficient time and information in decision analysis (Shen and Chung 2002; Shen *et al.* 2004). These problems are associated with the large number of participants and the short duration of VM workshops featured in Hong Kong. There is a strong demand for improvements to the practice in order to make the process more efficient and effective. Another two concerns come from the lack of VM knowledge among the participants and historical references for VM studies. First, since participants of VM workshops often do not know clearly what VM is, many facilitators start the workshop with a brief introduction to VM. It will be useful if a tool can be provided to introduce the generic process of VM studies before the workshop. Second, if an electrical database for VM studies can be set up, it will provide useful information for similar projects in the future.

The use of group support systems can overcome some of the difficulties frequently encountered among large teams (Dennis 1991). GSS is an interactive computer-based information system which combines the capabilities of communication technologies, database technologies, computer technologies and decision technologies to support the identification, analysis, formulation, evaluation and solution of semi-structured or unstructured problems by a group in a user-friendly computing environment (Er and Ng 1995). As a branch of information technology, GSS has the potential to promote active participation, encourage interaction and facilitate decision analysis in VM workshops. GSS can provide discussion support, information support, collaboration support and decision analysis support to VM workshops (Shen and Chung 2002).

This chapter introduces a research project which aims to investigate whether the use of GSS will enhance the efficiency and effectiveness of VM workshops. The research is not trying to replace face-to-face (FTF) mode with GSS but to integrate both modes of communication to exploit their full benefits. A web-based prototype GSS named Interactive Value Management System has been developed to support collaborative working in VM workshops. Two experimental validation studies have been conducted to investigate whether the use of IVMS can improve the performance of collaborative working in VM workshops.

2 Development of IVMS

A variety of GSS packages have been built by different university research teams and other organizations since the 1980s, and many GSS packages such as GroupSystems and Decision Explorer are available in the market. However, they typically offer a small set of tools such as electronic brainstorming and idea evaluation or voting to support discussion and decision. In order to overcome the problems listed above, to make the VM process more efficient and effective more specific functions are required, such as idea generation, function analysis and decision matrix. With the recent development of web technology, it is possible for Internet applications to address problems in various areas. For example, Palaneeswaran and Kumaraswamy (2005) presented a web-based client advisory decision support system for design–builder prequalification; Lee *et al.* (2006) introduced a web-based system dynamics model for error and change management on construction design and construction projects; Xie and Yapa (2006) developed a web-based system for large-scale environmental hydraulics problems. A web-based numerical model system has many advantages over a desktop model system, including easier access to distributed data and to the model system, efficient upgrades, improved compatibility, better user–developer communications, improved maintenance of security and the integrity of the model, and limiting access to protected data (Xie and Yapa 2006). Hence the authors integrated web technology and GSS to build a web-based GSS prototype named IVMS during this research to meet the needs of VM studies.

Purpose of IVMS

IVMS aims to supply a useful toolbox which can support VM practitioners in overcoming the problems in traditional VM workshops. Another concern is to build a computerized project database that contains various types of projects, which can be used as references for similar projects. It should be stressed again that IVMS is designed not to replace traditional VM procedures but to act as a beneficial complement by providing technical features. The system can be used by a team to integrate with the traditional face-to-face method to exploit the full benefits of both modes of communication. IVMS is designed to address the common problems frequently and to maximize the benefits of VM studies, as shown in Table 18.1.

Table 18.1 Proposed support by GSS in VM workshops

Problems	Reasons	Proposed support by GSS
Short duration	Pressure from the client to cut the cost.	Various electronic tools, including document library, electronic brainstorming, weighted evaluation tools, etc. to simplify and standardize the process.
Lack of information	Poorly organized project information in the pre-workshop phase. Difficulty of retrieving project information in meetings.	Information support such as document library, electronic discussion board and online questionnaire survey to improve the efficiency of information sharing and enhance information circulation.
Lack of participation and interaction	Shy about speaking in public. Dominance of a few individuals. Pressure to conform.	Virtual meeting rooms.
Difficulty in conducting analysis and evaluation	Insufficient time to complete analysis. Insufficient information to support analysis.	Electronic tools, including ideas categorizing and FAST diagram, etc. to improve the productivity and accuracy of data processing and eliminate human errors.
Database of VM studies	Provide references to similar projects in the future.	An electronic database that stores VM studies, including the process, the tools used, the objectives and outcomes, etc.
Lack of VM knowledge	Many participants not familiar with VM.	GSS can act as a teaching tool to introduce the generic process of VM.

Source: Shen *et al.* (2004).

Development environment

IVMS is built based on the Windows SharePoint Services (WSS) from Microsoft, which serves as a platform for application development. Microsoft Windows Server 2003 is adopted as the operating system (OS), and Microsoft SQL server, which supports concurrent data access, is adopted as database management system (DBMS). Microsoft Visual Studio .Net 2003 is used as the development environment of the application system. The system is coded mainly by using ASP.NET (Active Server Pages.NET), C# and JavaScript.

The purpose of IVMS is to help increase individual and team productivity. Including such IT resources as team workspaces, email, presence awareness and web-based conferencing, WSS enables users to locate distributed information quickly and efficiently, as well as connect to and work with others more productively. With the help of WSS, IVMS can be easily integrated with other useful software, including Microsoft Visio, Office and Messenger. Based on the functions provided by WSS, IVMS integrates GSS with the VM methodology to provide useful support to overcome problems in VM workshops.

Based on the characteristics of VM workshops and the features of GSS, the system structure of IVMS is designed, as shown in Figure 18.1.

3 Using IVMS in VM workshops

IVMS is designed to facilitate information management, improve communication and assist decision analysis in the VM workshops.

Although there are various application procedures at different stages of

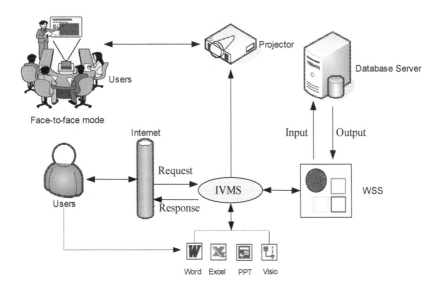

Figure 18.1 System architecture.

the project, all of them are more or less following the standard VM job plan. Male *et al.* (1998) gave a generic VM process in their benchmark study for value management. Figure 18.2 illustrates the process and outlines the steps.

The following section will discuss the use of IVMS in VM workshops in detail.

Pre-workshop phase

The pre-workshop phase provides an opportunity for all parties to understand project issues and constraints and therefore, gives and receives information before VM workshops. As mentioned previously, among the main problems in VM workshops are the poorly organized information and difficulty of retrieving project information in workshops, which are also suggested by Park (1993). In order to overcome these problems, several tools are provided. For example, a document library is provided for users for storing and sharing the project information. Its main features include: version history, permission management, check in/out mechanism, email alert, full-text search capability, and view selection.

A bulletin board is provided by the system for users to disseminate ideas and conduct discussions. One or more pre-meetings may be held in this phase to ensure that everyone involved in the project understands all the issues and constraints. Whenever the users have questions or fresh ideas, they can post their views on the bulletin board; others can read them and give replies or suggestions.

Questionnaire survey in this phase is usually used to collect the views of participants to give the facilitator an overview of strategic and tactical issues surrounding the project. The system provides an easy way to conduct online questionnaire surveys. The facilitator can make the participants answer the questions in different types, including rating scale or multiple-choice answers. If respondents' names are designed to be visible when the survey is set up, the users can see how each team member responded. The system also provides a 'graphical summary view' to display a compilation of responses.

An electronic agenda provides an outline of the workshop, including details of those who will lead the discussion on each subject, and the time that is allotted. This ensures that the team members are prepared.

When the workshop durations are driven shorter by the market, these features will improve the efficiency of information sharing and enhance the information circulation, and enable the facilitators to easily computerize and centralize the information gathering, distribution and circulation processes throughout VM workshops.

Information phase

In this phase, information relating to the project under review needs to be collected together, for example costs, quantities, drawings, specifications,

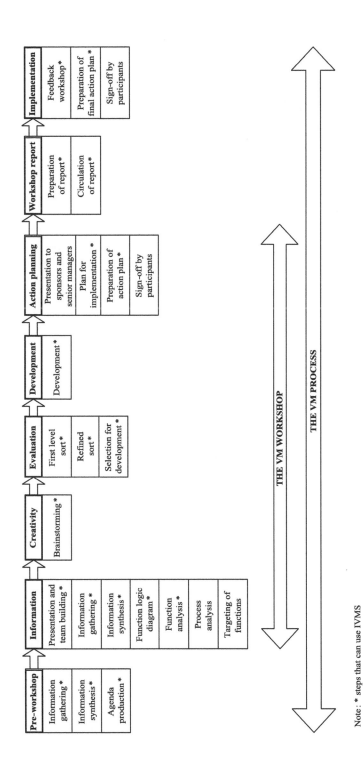

Note: * steps that can use IVMS

Figure 18.2 The generic VM process.

Source: Adapted from Male *et al.* (1998).

manufacturing methods, samples and prototypes (Kelly *et al.* 2004). The objective is to identify, in clear, unambiguous terms, the issues and functions of the whole or parts of the projects, as seen by the client organization (Male *et al.* 1998). This phase can be divided into two major parts: one is information sharing; the other is function analysis.

A VM workshop is commonly held in a conference room, a semi-closed environment with physical boundaries which may prohibit the users from retrieving any new information during workshops. The connection to the Internet breaks the physical boundaries of the conference room and allows members to access external information easily during the workshop. To enhance this web-based feature of this system, a database including various websites is provided to facilitate users in searching for information. The participants can also add new useful website links to the database to enrich it.

The 'document library' also plays an important part in the process of retrieving information, as all the information related to the VM workshops can be stored in the system before the workshop. The files, especially the files to be presented, can be shown on the large common viewing screen with the help of an LCD projector or 'public' screen at each member's terminal through the system, which makes the process of reviewing data more efficient.

Function analysis phase

The function analysis phase aims to clearly define the work involved and the requirements of the project (Assaf *et al.* 2000). Functional Analysis System Technique (FAST), developed by Charles W. Bytheway, is a standard VM tool which facilitates function analysis. The technique begins with a brainstorming session, which aims to generate functions required by the product or service. All functions are expressed as an active verb plus a descriptive noun. The functions generated are sorted by the VM team to format a diagram. Within the diagram, higher-level functions appear on the left hand and lower-level functions on the right hand. In this way, the logic relationships between the functions of the product or service will be systematically demonstrated. At the end of this phase, the functions which need to be improved will be selected for further studies in the next phases of the VM workshop.

The system provides support to the users in the brainstorming process of FAST. Virtual meeting rooms are provided to support users in the brainstorming session. According to different situations, the environment of these rooms can be switched between anonymous and nominal mode. In both modes, users can see all the functions that have been generated by others on their own computer, so users may spur each other on in generating functions. Further, the functions are stored automatically in the system as generated. Compared to recording the generated functions on paper in the traditional way, this feature can save much time.

Following the generation of functions, the VM team is invited to order the functions by putting the highest-order need at the far left side and the lowest-order want at the far right side. Some commonly used software, such as Microsoft Visio and Excel, can be used to integrate with the system to provide model tools. The LCD projector can be used to display the data analysis process.

Creativity phase

The main task of this session is to generate numerous alternatives for accomplishing basic functions required by the clients, by means of creativity-stimulating techniques, such as brainstorming, synectics, morphological charts and lateral thinking (Shen and Shen 1999). Brainstorming is the most popular technique in the creativity phase. It requires that users consider a function and contribute any suggestions which expand, clarify or answer that function. However, some participants are reluctant to speak out in this phase because they are shy of speaking in public or afraid of being criticized or sounding stupid (Lamm and Trommsdorf 1973; Diehl and Stroebe 1987; Mullen *et al.* 1991; Camacho and Paulus 1995). Moreover, this process can be dominated by a few individuals, which can make the creativity process very unproductive.

In order to overcome these communication problems, the system provides virtual meeting rooms, which are like the chat rooms currently popular on the Internet. One of the basic rules of the brainstorming process is that the group should be relatively small (e.g. up to eight members) (Norton and McElligott 1995). However, there can be 20–30 stakeholders in a VM work-shop, in which case workshop members are nominated to five 'rooms' in the system. Workshop members 'go' to the assigned room and type their ideas and submit them under the special functions which have been chosen in the 'function analysis' phase. As shown in Figure 18.3, functions are shown at the top of the interface to make them obvious to participants. The function can be changed by the facilitator. The left part shows the names of members in this room. Each member can read on his/her own screen the ideas generated by others.

The following paragraphs introduce the main features of the virtual rooms which are designed to make the brainstorming session more effective and efficient:

- *Flexible anonymous or nominal mode in idea generation.* The environment can be set to be totally anonymous or nominal according to the need of the workshops. When the environment is anonymous, each user can read on his/her screen the ideas generated by other group members without knowing from whom they originate. Users who fear receiving negative comments from others in the face-to-face session may appreciate the environment of anonymity in IVMS. This form of anonymity can

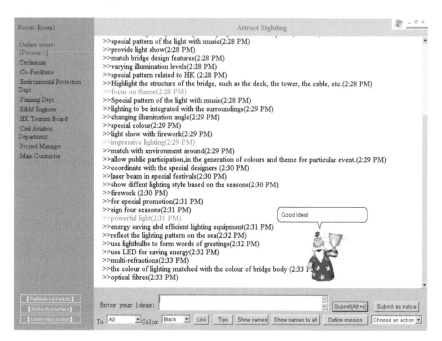

Figure 18.3 Virtual meeting rooms.

reduce evaluation apprehension losses (Connolly *et al.* 1990; Gallupe *et al.* 1991, 1992). However, it does not mean that the nominal environment should not be used. While an anonymous environment encourages participants to express their ideas freely, it may also lead to a situation where some work hard and some free-ride on the efforts of others. In a nominal environment, the users' names are displayed with the ideas they generated, giving them the stimulus to generate more ideas to prove themselves. Hence the system provides the opportunity to choose the environment mode flexibly to exploit the full benefits.

- *Brainstorming agent.* It is found that there is more task-focused communication and less joking and laughing in GSS-supported groups (Turoff and Hiltz 1982) and people are more critical of each other's ideas when they communicate electronically (Siegel *et al.* 1986). DeSanctis and Gallupe (1987) also suggest that features intended to address the social needs of groups should be included in GSS systems. The IVMS provides an agent that can pop up with different words and gestures corresponding to the situation. The agent can 'monitor' the performance both of the whole group and of individuals. The agent measures the performance of the whole group based on several criteria, including the quantity of ideas generated by the group, the idea generation rate, and the percentage of participants in the group who are active in generating ideas. The agent will give the participants applause when they generate ideas actively. An individual's performance is monitored by the agent according to the quantity of ideas he/she generated, the idea generation rate and the time during which he/she kept silent. The system provides an alternative way to maintain active participation of all participants. When users are criticized by an animation agent, they may feel less embarrassed.

- *Control functions for the facilitator.* The control functions are the functions that only the facilitator can use in the VM workshops, including changing the environment mode, editing/deleting unnecessary ideas, posting VM notices and so on. This setting makes it convenient for the facilitator to control the whole process. For example, if someone in the group disrupts the brainstorming, the facilitator could put out some notice or criticize him/her secretly through the agent to make the workshop go smoothly. The facilitator in a traditional workshop can only encourage publicly, while IVMS allows anonymous encouragement.

- *Tips for generating more ideas.* This function is designed to inspire the users by providing some constructive suggestions, for example 'What if ice cream were hot?' or 'What if pigs could fly?' The aim is to provide 'triggers' to make the participants think in a different way, and fresh ideas may come out. Hence it should be considered to be useful even if only one tip gives the users some illumination.

- *Enabled parallelism in idea generation.* Parallelism helps reduce production blocking, since users no longer have to wait for others to express their ideas (Jessup *et al.* 1990; Gallupe *et al.* 1991). Users can express their ideas freely.

- *Other useful facilities.* Users can select different colours to display their ideas, which could make the ideas more attractive and easily distinguished from those of others. Internet links also can be posted during the brainstorming process.

Using the system, the participants can conduct the brainstorming to generate ideas or issues before the workshop, which can shorten the duration of a workshop. Moreover, all the stakeholders can take part in the brainstorming by using IVMS, compared to only selected stakeholders being able to join the brainstorming in traditional workshops.

Evaluation phase

According to the Value Methodology Standard (SAVE International 1998), the main tasks in this phase include setting up a number of criteria and evaluating and selecting alternatives generated during the creativity phase. Various models and techniques, such as cost models, energy models and the weighted evaluation technique (WET), are used during this phase. In addition, some form of weighted vote is also used (Kelly *et al.* 2004). This system provides an electronic weighted evaluation technique.

Idea categorization. Ideas generated in the 'creativity' session will be automatically collected and listed corresponding to the functions. The facilitator can delete the overlap ideas and correct grammar or spelling mistakes. These ideas will then be classified by the workshop participants into P1 – 'Realistically possible', P2 – 'Remotely possible' and P3 – 'Fantasy'. Only P1 ideas will be considered further in the subsequent phases.

Weighted evaluation. This step includes three tasks: list criteria, assign weighting and score ideas. They are explained as follows:

- *Step 1: List criteria.* This part facilitates members in setting up a number of meaningful criteria, against which the P1 ideas can be evaluated. It is important not to select criteria which are highly correlated with each other. Since one set of criteria may be used more than once, or be very similar to another set, the system provides a function called 'Criteria Database'. When one set of criteria has been set up, the criteria can be saved as a template. If the users want to set up a similar set of criteria or use this set again, they can load the template.
- *Step 2: Assign weighting.* Since the criteria are not of the same importance, a relative importance weighting is assigned to each of the established criteria. The system provides an electronic team-oriented pair-wise method to determine the weighting to be given to each criterion. The system assigns each criterion with a letter of the English alphabet. The preferences are selected from a pop-up list with two parts of fixed entries. The first part is the letters of the two criteria which are being compared. For example, if criteria A and B are being compared, the entries of the first part will be fixed to A and B. The other part is four fixed entries: 1 means 'Slight, no preference', 2 means 'Minor preference', 3 means 'Medium preference', and 4 means 'Major preference'. For example, if A is much more important than B, a score of 4 will be assigned to criterion A during the comparison, that is, A/4. This is repeated for each pair of criteria. The system will automatically calculate the final scores as the sum of all the numbers in every score, ΣAx. Figure 18.4 is the 'Scoring Matrix' screen. The process can be repeated if necessary.
- *Step 3: Score ideas.* An interactive table is generated corresponding to the quantity of criteria and the P1 ideas from the previous phases. Participants need to score every idea under each criterion. Each score is multiplied by the criterion's weighting, and the subtotal will be automatically computerized as the final score of each idea. To focus the participants, only one criterion is displayed when scoring ideas; IVMS hides other criteria. Figure 18.5 shows the screens of 'Score Ideas' and the outcome. The process can be repeated to revise the outcomes, for example when there is a divergence among the participants.

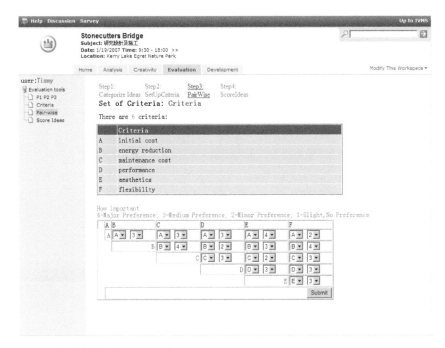

Figure 18.4 Pair-wise comparison.

Development phase

The development phase investigates selected alternatives in sufficient depth, such that they can be written into recommendations for implementation. This involves not only detailed technical and economic evaluation but also consideration of the applicability for implementation (Shen and Shen 1999). There is wide scope for the use of lifecycle cost models and computer-aided calculations at this stage (Kelly *et al.* 2004). A whole lifecycle cost toolkit has also been developed using Visual Basic for Applications (VBA) for Microsoft Excel to support VM at this particular stage.

In this phase, the team defines and quantifies results, and prepares and presents recommendations to the decision makers. It includes presentation to sponsors and senior managers, a plan for implementation, preparation of an action plan, and sign-off by the participants. The document library can be used to make the presentation process more flexible and effective. Users can upload the files to the document library, and others can view them. When subgroups have been formed to develop ideas, the system is more useful, for the reports of each group can be collected and shared quickly through the document library. The system also facilitates the preparation of an action plan, by providing task management functionality such as Gantt charts for visualization of task relationships and status. The function of email

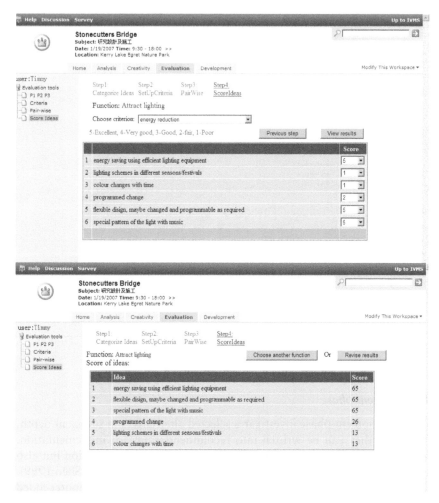

Figure 18.5 Idea evaluation.

notification is integrated, and a notification will be sent to the members when they are assigned a task.

Reporting and implementation phase

A detailed report must be prepared as soon as possible after the workshop. The workshop report must then be circulated to the workshop participants to confirm their role in the implementation of the workshop proposals and any further development work necessary (Male *et al.* 1998). Since the system can automatically collect the main information of the workshop, such as the quantity of the ideas, the time of each phase of the workshop and so on, it will be helpful to the preparation of the workshop report. The document library can also be used to circulate the report. The information collected could be used to evaluate the processes and outcomes of the workshop. There is also a questionnaire at the end of the workshop to collect the users' views on the workshop. The information collected by the system and the questionnaire will produce a score for the workshop. Although this score cannot tell participants exactly whether this VM study is good or bad, it can give the users a general picture on workshop performance.

The objective of implementation is to ensure the proper implementation of the approved value study change recommendations and collect feedback on the proposal. The system provides several efficient ways to collect feedback, such as a bulletin board, questionnaires and a notice board. Through these functions, users can conduct online discussions, post their ideas, and submit their feedback whenever they like through the Internet.

As introduced previously, we plan to build a computerized project database which can be used as a reference for similar projects. All the information involved during the whole VM process will be automatically stored in the database. The users can search related workshops by the title, date, facilitator's name or keywords for reference.

4 Validation of IVMS

In building a computerized conferencing system, Hiltz and Turoff (1981) found that 'Users cannot tell you what they need prior to using the technology.' Consequently, users must have extended experience with GSS before the system design can be fully assessed (DeSanctis and Gallupe 1987). Two experimental studies have been designed and conducted in order to validate the IVMS.

A GSS room was established to conduct GSS-supported VM workshops during these two studies, as shown in Figure 18.6. Each participant was provided with a laptop, and all the laptops are linked together through a wireless network. Each user can access the system on his/her laptop during the whole-day workshop. A projector and a large common viewing screen were also provided in the GSS room, which can display the public notice or the group information. The movable seats and tables can be arranged in a U-shape or a semi-circle for different situations.

Experimental study 1 – a comparison study of idea generation

The first experiment conducted was a comparative study of idea generation between VM workshops with IVMS support and VM workshops without IVMS support. This study aimed to validate the communication support of IVMS. The participants were formed by 72 students enrolled in a VM course in the Hong Kong Polytechnic University. The subjects were divided randomly into 12 six-person groups to mirror the size of the discussion groups in real-life VM workshops in Hong Kong. Six groups used IVMS-supported

Figure 18.6 VM study with IVMS support.

brainstorming, and the others used face-to-face brainstorming to generate ideas for the same task, which was extracted from a real project in Hong Kong. The detail design of the experiment is shown in Table 18.2. In order to control the impact of facilitator styles, the same facilitator was invited to facilitate the brainstorming process. Hence the experiment was divided into two one-day sessions: six groups conducted face-to-face brainstorming on day 1; the other six groups were provided with IVMS support on the next day. Six researchers acted as observers to record useful information such as the number of ideas generated by individuals in the face-to-face brainstorming process and the typing speed of participants in the IVMS-supported brainstorming process.

The ideas generated were compared according to the quantity and the quality. Quality was assessed by two researchers, who categorized the ideas

Table 18.2 Framework for the proposed experimental studies

Session	Group	Brainstorming approach
Session A	Groups 1–6	Face-to-face brainstorming: participants speak out their ideas one by one, and the ideas are recorded at the same time by a recorder.
Session B	Groups 7–12	IVMS-supported brainstorming: participants generate ideas in an anonymous environment through IVMS, and the ideas are recorded automatically by the system.

into P1 – 'Realistically possible', P2 – 'Remotely possible' and P3 – 'Fantasy'. In this way, P1 ideas were considered as the ideas with good quality. The ideas generated in each group were collected, and repeated ideas were removed. Then the unique ideas were categorized by the two researchers. The quantity of unique ideas and unique ideas with good quality (equal to P1 ideas) are shown in Table 18.3.

The quantity of unique ideas

Table 18.3 shows that the average of unique ideas generated by FTF groups was less than the average of unique ideas generated by IVMS groups. Moreover, there is a significant difference between the two sets of data (significance $p=0.03<0.05$), which means that the number of unique ideas generated through IVMS was statistically larger than that through the face-to-face approach. One of the most important reasons for this phenomenon is the parallel entry of ideas, that is, using IVMS users did not need to wait their turn to express their ideas. In contrast, only one user at a time could express ideas in the face-to-face mode. The environment of total anonymity of IVMS is another important factor. The results of the observation showed that participants in an IVMS session were more active than the ones in the face-to-face session. It is the factor of anonymity that caused the users to become more active. Users who fear receiving negative evaluations from others in the face-to-face brainstorming do not have this fear in the environment of anonymity in the IVMS-supported process.

The quality of ideas

The quality of ideas was measured by three variables: 1) the quantity of unique ideas with good quality, 2) the width of ideas, and 3) the depth

Table 18.3 Quantity of unique ideas and unique P1 ideas

	Unique ideas		Unique P1 ideas	
	Face-to-face (Groups 1–6)	IVMS (Groups 7–12)	Face-to-face (Groups 1–6)	IVMS (Groups 7–12)
	20	44	8	17
	25	60	15	11
	31	50	13	15
	23	31	14	21
	21	25	12	14
	23	24	10	19
Total	143	234	72	97
Mean	24	39	12	16
t-test		p=0.03*		p=0.04*

Note: * means p<0.05.

of ideas. As mentioned above, the unique ideas with good quality were selected by two researchers. As Table 18.3 shows, the average of P1 ideas generated by the IVMS-supported groups was larger than the average P1 ideas generated by the face-to-face groups. The results of the t-test also show that there was a significant difference between the two sets of data (p=0.04<0.05), which means that the number of unique P1 ideas generated through IVMS was significantly larger than that through the face-to-face approach.

Then an idea tree was developed to analyse the width and depth of ideas generated. The ideas generated by all the groups were collected together, and then repeated ideas were removed. Hence, although there were 377 ideas (FTF: 143; IVMS: 234) generated according to Table 18.3, there were 236 unique ideas left after the removal of repeated ideas. These unique ideas were further divided into five divisions and 24 branches. The ideas of each group were compared with the idea tree. When the width of ideas was analysed, only the divisions were considered. If an idea related to one of the branches, a 100-score would be calculated. For example, if the ideas generated from one group related to four divisions, the width was 400. When the depth of ideas was analysed, the branches were considered. Each branch mentioned by the ideas was calculated as a 50-score. The width and depth of ideas are shown in Table 18.4.

From Table 18.4, it can be found that the width of ideas generated using GSS was nearly the same as the width of ideas generated by the face-to-face groups, which is also supported by the t-test (p=0.14>0.05). The results of the t-test also show that there is no significant difference between the ideas generated through the two approaches in the depth of ideas, with everything else remaining equal. The results suggest that the depth and width of ideas do not change with the brainstorming approaches in VM workshops. It also seems that GSS is important for quantity but not for quality. However, a possible reason for this phenomenon is that the tasks used were all simple ones and related to the students' study; hence it is not surprising that they

Table 18.4 Width and depth of ideas

	Width of ideas		Depth of ideas	
	Face-to-face	GSS	Face-to-face	GSS
Group 1	400	500	200	300
Group 2	400	400	450	250
Group 3	400	500	200	400
Group 4	400	400	400	500
Group 5	400	400	300	400
Group 6	400	400	350	300
Mean	400	433	317	358
t-test		p=0.14		p=0.48

generated ideas with similar width and depth. Further investigation should be conducted to investigate the effect of GSS on the width and depth of ideas on more difficult tasks.

A limitation of this study is the subjects. There is often a concern raised regarding the use of students as subjects in GSS research. Fjermestad and Hiltz (1999) report that 94 per cent of those studies involved students as subjects. The limitation of using students as participants had been recognized long before (Lorge *et al.* 1958), but there are still many researches using students owing to the difficulty in persuading real managers to participate in GSS sessions. However, the authors think that it is useful to use students as subjects as the first step of system validation. It is much easier to invite students, compared with real-life practitioners, to participate. Referring to the outcomes, Briggs *et al.* (1996) found no significant differences between executive business managers and graduate business students in evaluating technology. Also, Remus (1986) found no significant differences between line managers and MBA students with little business experience in production scheduling decisions. In the light of these findings, we felt comfortable with the background of our participants.

Experimental study 2 – using IVMS in the full process of VM workshops

This study was conducted in April 2005 to investigate the effect of using IVMS in the full process of VM workshops. A real project task, Stonecutters Bridge in Hong Kong, was taken as the object. There were two main tasks in this study, which were to review the colour of the architectural lighting for the bridge and construct a look-out point and exhibition centre (LOP & EC) for the bridge. The objective of workshop No. 1 was to recommend architecture lighting themes and colours to be adopted on the bridge at different times of the year without any implications to the structural detailing. The objective of workshop No. 2 was to identify the functions for the LOP & EC and a layout for parking facilities, and decide whether a café facility and other leisure facilities should be provided based on the endorsed location beside the western tower of the bridge.

The participants were 20 practitioners, all of whom had been working in the construction industry for at least several years. During this VM study, they represented the necessary stakeholders relevant to the task, including: client, designer, quantity surveyor, government representatives and other stakeholders. Since all of the participants were now working in real projects, their work experience would make this VM study just like a real-life VM study.

The participants were asked to organize a GSS-based VM workshop with the whole process, including pre-workshop phase, workshop phase and post-workshop phase. All of the participants undertook training to make sure that none of them had any problems in using the system. The workshop was facilitated with IVMS support. The system was used exactly according to section 3 of this chapter, 'Using IVMS in VM Workshops'.

In order to test IVMS, a questionnaire survey was conducted during this study. Each item was measured on a five-point Likert-type scale, with 1 as 'Strongly disagree' and 5 as 'Strongly agree'. Five open-ended questions were also included in the survey to collect the suggestions from the participants on IVMS.

The findings of the VM study suggested that the application of IVMS in VM workshops was similar to its application with students. Table 18.5 shows that most of the IVMS functions were reported to be useful in supporting and improving VM workshops; the scores of 9 in 12 items are not less than 4.00, and 3 items more than 3.00. This provides strong evidence to support the idea of using GSS tools to improve VM workshops.

The results also show that the statements 'IVMS can promote active participation in idea generation' and 'IVMS can avoid conformance pressure in idea generation' were the ones that the participants agreed with most. As mentioned previously, conformance pressure and lack of active participation are the main problems in VM workshops. It can be concluded from the survey that IVMS can provide strong support in overcoming these problems, which is also supported by the participants' answers to the open-ended questions. When the participants were asked 'What are the things that you like most about IVMS?', the encouraging environment for discussion and creativity was most frequently mentioned. The staged outcomes of the creativity phase in this study were another indication. There were 147 ideas

Table 18.5 Summary of the survey results on the support of IVMS

Type of support	Average
Support in information phase:	
IVMS can improve the availability of information.	4.18
IVMS can improve the information exchange process.	4.18
Support in function analysis phase:	
IVMS can simplify the function analysis processes.	4.12
IVMS can enhance the function analysis processes.	4.18
Support in creativity phase:	
IVMS can promote active participation in idea generation.	4.29
IVMS can avoid conformance pressure in idea generation.	4.24
IVMS can prevent domination in discussion.	3.94
The pop-up character in IVMS can enhance the atmosphere of creativity.	3.88
The function of 'Tips' can help me in generating ideas.	3.47
Support in evaluation phase:	
IVMS can simplify the evaluation processes.	4.06
IVMS can enhance the evaluation processes.	4.00
Interface of IVMS:	
I feel comfortable with the current interface of IVMS	4.00

5 Strongly agree; 4 Agree; 3 Neutral; 2 Disagree; 1 Strongly disagree.

generated in a 30-minute brainstorming session, which is regarded as further evidence for the system's effectiveness.

Improving the availability of information and the information exchange process both are ranked as the second most useful functions. This means that IVMS could help improve the lack of information in VM workshops. Information support was most appreciated when the participants were divided into several subgroups to conduct discussions on different topics. After the discussion, the outcomes of each subgroup were collected and disseminated to others. Traditionally, the outcomes were first collected by a coordinator or facilitator and then typed into a computer in order to show them to all the participants. During such a process, there must be a break to collect and tidy the outcomes, costing time and breaking the rhythm of discussion. In this study, each subgroup submitted the outcome to the system and then all of the participants could view it through the system immediately. In this way, time and repetitive typing work were saved. This is one of the reasons why participants recognize the information support provided by the system. Besides, the support in analysis and evaluation phases both achieve the participants' consent. The interface of the IVMS is important for its application. If the participants do not like the interface, the performance of the VM workshop will be influenced. However, the score of 'I feel comfortable with the current interface of IVMS' was 4.0, as shown in Table 18.5, which indicates that most of the participants feel comfortable with the interface of IVMS.

Above all, the results of experimental study 1 suggest that groups using GSS can generate more unique ideas and more unique good ideas than the groups with face-to-face brainstorming. Experimental study 2 indicates that the participants are satisfied with most of the support provided by IVMS, and agree that the IVMS can overcome the problems in VM workshops.

However, the system also has some limitations. Through the observation of researchers and the feedback of the participants, it is found that several aspects of the system need improving. Some participants suggest that a visualizing tool, for example an electronic whiteboard, should be added to the system. With it, the function diagram can be drawn by easily dragging and dropping and interactively visualized. A voting tool is also proposed by the users. The voting tool would provide the participants with a chance to express their own ideas during the workshop. The comments mainly focus on the guidance of the system. The uses suggest that the guidance of the system is too complex to follow. The system should be developed as simply as possible. These suggestions will help the authors to improve the system in future studies.

Computer skills, especially typing skills, should also be an important factor to be considered in future. Most participants who took part in the experiment had good computer skills. However, the participants in a real-life VM workshop who are senior members in companies or high-level professionals may not have good computer skills. In this case, the performance of the system would be affected. Hence, team members' computer skills and their effect on the team performance is an issue that can be explored in the further research.

5 Conclusions

The research described in this chapter is unique in its investigation of the effectiveness of using GSS in VM workshops. Although the two action research studies are limited to a laboratory setting, they influence the direction of future field studies. The results show that web-based GSS can improve the efficiency and effectiveness of VM workshops by supporting the VM team. The result of the validation of the system shows that IVMS is a useful tool in facilitating the information exchange process, encouraging interaction, and promoting active participation in VM workshops. It also reveals that web-based GSS can overcome some of the common problems identified in VM workshops, for example lack of information, short duration, and lack of participation and interaction. Focus group meetings with VM practitioners and real-life VM workshops are required in the further research to validate IVMS. A set of critical successful factors on the use of GSS in VM workshops will also be identified.

Acknowledgement

The authors wish to express their sincere gratitude to the Research Grants Council of the Hong Kong Special Administrative Region, China (PolyU 5114/03E and PolyU 5161/04E) for the funding support to the projects on which this chapter is based.

References

Assaf, S., Jannadi, O. and Al-Tamimi, A. (2000). 'Computerized System for Application of Value Engineering Methodology', *Journal of Computing in Civil Engineering*, 14(3), pp. 206–214.

Briggs, R. O., Balthazard, P. and Dennis, A. R. (1996). 'Graduate Business Students as Surrogates for Executives in the Evaluation of Technology', *Journal of End-User Computing*, 8(4), pp. 11–17.

Camacho, L. M. and Paulus, P. B. (1995). 'The Role of Social Anxiousness in Group Brainstorming', *Journal of Personality and Social Psychology*, 68(6), pp. 1071–1080.

Connolly, T., Jessup, L. M., and Valacich, J. S. (1990). 'Effects of Anonymity and Evaluative Tone on Idea Generation in Computer Mediated Groups', *Management Science*, 36(6), pp. 689–703.

Construction Industry Review Committee (2001). *Construct for Excellence: Report of the Construction Industry Review Committee*. Hong Kong: Printing Department, Hong Kong Special Administrative Region Government.

Dennis, A. R. (1991). 'Parallelism, Anonymity, Structure and Group Size in Electronic Meetings', Doctoral dissertation, University of Arizona, Tucson.

DeSanctis, G. and Gallupe, R. B. (1987). 'A Foundation for the Study of Group Decision Support Systems', *Management Science*, 33(5), pp. 589–609.

Diehl, M. and Stroebe, W. (1987). 'Productivity Loss in Brainstorming Groups: Toward the Solution of a Riddle', *Journal of Personality and Social Psychology*, 53(3), pp. 497–509.

Environment, Transport and Works Bureau (ETWB) (2002). *Technical Circular No. 35/2002: Implementation of Value Management in Public Works Projects*. Hong Kong: Government of the Hong Kong Special Administrative Region.

Er, M. C., and Ng, A. C. (1995). 'The Anonymity and Proximity Factors in Group Decision Support Systems', *Decision Support Systems*, 14(1), pp. 75–83.

Fjermestad, J. and Hiltz, S. R. (1999). 'An Assessment of Group Support Systems Experimental Research Methodology and Results', *Journal of Management Information Systems*, 15(3), pp. 7–149.

Gallupe, R. B., Bastianutti, L. and Cooper, W. H. (1991). 'Unblocking Brainstorms', *Journal of Applied Psychology*, 76(1), pp. 137–142.

Gallupe, R. B., Dennis, A. R., Cooper, W., Valacich, J. S., Bastianutti, L. M. and Nunamaker, J. F. (1992). 'Electronic Brainstorming and Group Size', *Academy of Management Journal*, 35(2), pp. 350–369.

Hiltz, S. R. and Turoff, M. (1981). 'The Evolution of User Behavior in a Computerized Conferencing System', *Communication of the ACM*, 24(11), pp. 739–751.

Jessup, L. M., Connolly, T. and Tansik, D. A. (1990). 'Toward a Theory of Automated Group Work: The De-individuating Effects of Anonymity', *Small Group Research*, 21(3), pp. 333–348.

Kelly, J., Male, S. and Graham, D. (2004). *Value Management of Construction Projects*. Oxford: Blackwell Science.

Lamm, H. and Trommsdorf, G. (1973). 'Group versus Individual Performance on Tasks Requiring Ideational Proficiency (Brainstorming): A Review', *European Journal of Social Psychology*, 3(4), pp. 361–387.

Lee, S. H., Peña-Mora, F. and Park, M. (2006). 'Web-Enabled System Dynamics Model for Error and Change Management on Concurrent Design and Construction Projects', *Journal of Computing in Civil Engineering*, 20(4), pp. 290–300.

Lorge, I., Fox, D., Davitz, J. and Brenner, M. A. (1958). 'A Survey of Studies Contrasting the Quality of Group Performance and Individual Performance, 1920–1957', *Psychological Bulletin*, 55(6), pp. 337–372.

Male, S., Kelly, J., Fernie, S., Gronqvist, M. and Bowles, G. (1998). *The Value Management Benchmark: A Good Practice Framework for Clients and Practitioners*, Report for the EPSRC IMI contract. London: Thomas Telford.

Mullen, B., Johnson, C. and Salas, E. (1991). 'Productivity Loss in Brainstorming Groups: A Meta-analytic Integration', *Basic and Applied Social Psychology*, 12(1), pp. 3–23.

Norton, B. R. and McElligott, C. W. (1995). *Value Management in Construction: A Practical Guide*, pp. 79–153. London: Macmillan.

Nunamaker, J. F., Vogel, D. R. and Potter, R. (1997). 'Individual and Team Trends and Implications for Business Firms', *Advances in the Study of Entrepreneurship, Innovation, and Economic Growth*, 9, pp. 199–247. JAI Press.

Palaneeswaran, E. and Kumaraswamy, M. M. (2005). 'Web-Based Client Advisory

Decision Support System for Design–Builder Prequalification', *Journal of Computing in Civil Engineering*, 19(1), pp. 69–82.

Park, P. E. (1993). 'Creativity and Value Engineering Teams', in *Proceedings of the 28th SAVE International Annual Conference*. Fort Lauderdale, FL: SAVE International.

Remus, W. (1986). 'Graduate Students as Surrogates for Managers in Experiments on Business Decision Making', *Journal of Business Research*, 14(1), pp. 19–25.

SAVE International (1998). *Value Methodology Standard*, 2nd rev. edn, pp. 4–18. Northbrook, IL: SAVE International.

Shen, Q. P. and Kwok, E. (1999). 'The Economic Downturn in Hong Kong: Crisis or Opportunity for Sustainable VM Applications?', in *Proceedings of the 3rd International VM Conference*, pp. 6–7. Hong Kong: Hong Kong Institute of Value Management.

Shen, Q. P., and Shen, L. Y. (1999). 'Value Management as a Vehicle for Scope Management of Construction Projects', *Journal of Harbin University of Civil Engineering and Architecture*, 32(5), pp. 107–115.

Shen, Q. P. and Chung, K. H. (2002). 'A Group Decision Support System for Value Management Studies in the Construction Industry', *International Journal of Project Management*, 20(3), pp. 247–252.

Shen, Q. P., Chung, K. H., Li, H. and Shen, L. Y. (2004). 'A Group System for Improving Value Management Studies in Construction', *Automation in Construction*, 13, pp. 209–224.

Siegel, J. V., Kiesler, S. and McGuire, T. W. (1986). 'Group Processes in Computer-Mediated Communication', *Organizational Behavior and Human Decision Processes*, 37(2), pp. 157–187.

Turoff, M. and Hiltz, S. R. (1982). 'Computer Support for Group versus Individual Decisions', *IEEE Transactions on Communications*, 30(1), pp. 82–90.

Works Bureau (1998). *Technical Circular No. 16/98, Planning, Environment and Lands Bureau Technical Circular No. 9/98: Implementation of Value Management*. Hong Kong: Government of the Hong Kong Special Administrative Region.

Xie, H. and Yapa, P. D. (2006). 'Developing a Web-Based System for Large-Scale Environmental Hydraulics Problems with Application to Oil Spill Modeling', *Journal of Computing in Civil Engineering*, 20(3), pp. 197–209.

Index